British Sheep and Shepherding

by Walter James Malden

with an introduction by Jackson Chambers

Self Reliance Books

Get more historic titles on animal and stock breeding, gardening and old fashioned skills by visiting us at:

http://selfreliancebooks.blogspot.com/

Introduction

I am pleased to present yet another practical title on breeding and raising livestock.

The work is in the Public Domain and is re-printed here in accordance with Federal Laws.

As with all reprinted books of this age that are intended to perfectly reproduce the original edition, considerable pains and effort had to be undertaken to correct fading and sometimes outright damage to existing proofs of this title. At times, this task is quite monumental, requiring an almost total "rebuilding" of some pages from digital proofs of multiple copies. Despite this, imperfections still sometimes exist in the final proof and may detract from the visual appearance of the text.

I hope you enjoy reading this book as much as I enjoyed making it available to readers again.

Jackson Chambers

PREFACE.

Britain is the home of sheep breeds and the breeding place of skilled flock-masters. There are soils, climates, and indigenous stock, together with a wide range of husbandries, that enable the breed-maker to build up types suitable to his home needs and the world's requirements. Whilst this is so, the subject of sheep must have wide interest, and no apology is needed for bringing out a modern book on sheep. It was written just before the outbreak of the great war. Sheep have become very dear, and all kinds sell readily—even the larger breeds. With prospective high prices for a long time, some remarks may appear to be of less force than they would have been, but some of the principles I have advocated are only temporarily affected, and time will readjust them. Much that was set out in my previous book, "Sheep-raising and Shepherding," has been incorporated, in slightly altered form, because many teachers of agriculture desired that it should be, as it was convenient for their work.

The general object has been to produce a book both practical and suggestive. Personal experience in handling many breeds in their native districts, and many when away from them, enables me to write with a confidence I otherwise could not presume. Although I have handled many Scotch sheep, I have not lived on their native hills; therefore I have to thank Mr. Anderson who has had much experience with them on Scotch hills, and who is now successfully engaged in developing new breeds in Wales, for the reliable record he has set out. Mr. A. Mansell and others among flock improvers, as well as those associated with many flock books, have supplied information as I have needed it, and I gratefully acknowledge their kindness. Captain Leeney, A.V.C. (then Mr. Harold Leeney, F.R.C.V.S.), wrote an article in the Journal of the R.A.S.E. on "The Lambing Pen," and this in pamphlet form has been sold in vast numbers. Undoubtedly hygienic principles at lambing time as a common practice date from the publication of this paper, and thousands of ewes and lambs are saved yearly as a result. I felt it would be greatly appreciated if the Veterinary Section of this book were written by Captain Leeney, as he could treat it with an accuracy and up-to-dateness which a layman would not do; so, with the co-operation of the publishers, this was arranged. That it has added to the value of the book there is no doubt, and I welcome it with much pleasure.

As I have not an expert knowledge of wool when it has left the sheep, Professor Barker, of the Leeds Technical College, kindly undertook to write a short section, which is full of interest and instruction.

I would like to call attention to the fact that Chapters XI. to XIX. are specially designed to enable the inexperienced to follow out the work with the different sections of the flock day by day, and season by season, throughout the year. It sets out what the farmer actually has to do, and in this form I have been assured on many occasions that the year's shepherding and sheep management can be readily grasped by those whose experience is very limited. I would, therefore, suggest that these chapters be read consecutively, so that the thread of the work be gripped and intelligently understood. A mass of facts without connectiveness rarely proves useful; and knowledge itself is of little use unless it is turned to account; and in writing simply I have tried to reduce to practice anything that has been put forward, so that the least experienced may understand. I would urge the ancient adage: "With all thy getting, get understanding."

To those who are devoting time to the making and improving of breeds, all owe thanks; doubtless they find in it ample interest and reward, for it is a life full of interest.

W. J. MALDEN.

CONTENTS.

CONTENTS.

CONTENTS.

ILLUSTRATIONS.

INTRODUCTION.

THE shepherd and the flockowner have to work hand in hand. No man on the farm carries so much responsibility as the shepherd, and no man on the farm is so self-sacrificing of his ease and strength as the shepherd in the interest of sheep under his care. In some large sheep-farming districts the shepherd is scarcely second to the farmer, for he often dictates the cropping, and the time for consuming it, as well as the time for selling the sheep. In other instances the farmer takes most control, but then a skilled and trustworthy shepherd is needed—for sheep require an eye on them at all times.

Shepherds are generally shepherds by descent; this ensures an instinctive love and appreciation of sheep, as well as a very considerable amount of sheep-lore which has passed from mouth to mouth through generations. A very large proportion of this knowledge is highly valuable, but some of it is not reliable; wrong ideas prevail in shepherding as in other things, and the good and the bad have come down together. It was not until comparatively recent years that sheep came under the influence of really progressive science; some superficial scientific knowledge was brought to bear on the sheep, but few men of science devoted their time to the study, because there was little prospect of profit in their doing so. In more recent years, however, the sheep has had its full share of attention from the scientist, and with marked benefit. There is much more to be done, and that, naturally, the more intricate and submerged. Still there is a great field for the man of research, and it may with truth be said that in discovering preventive and curative means, reasonable to apply, in respect to some of the diseases affecting sheep, there is every inducement that pecuniary reward can give. There are diseases which levy a heavy toll from sheep-keepers throughout the world, and the man who makes discoveries to substantially hold them in check has a very great prospect of wealth before him. This is in great contrast to the days, not so very long ago, when sheep medicine and surgery were regarded lightly and so little appreciated that it was scarcely worth while for a veterinary surgeon to devote time to acquiring special knowledge in relation to sheep.

Mention has been made of the knowledge brought down by tradition. Within recent years what are undoubtedly very old diseases or affections, some not even recognised, others surmised, but in no way understood, have been brought under skilled observation, often with the result that they are now kept well in check. In other cases the cause has been discovered, but at present the remedy is awaited, though vigilant experimenters are constantly at work to find a cure.

There are two diseases, especially, which have been brought to light, and which await defeat—the strongyle worms in the fourth stomach, and Johne's disease. It is difficult to express how much these diseases have influenced sheep-raising, but owing to the want of recognition of their existence, many very strange things have been perpetrated in shepherding, and very ridiculous conclusions have been drawn as to the influence of matters which could not be guilty of the wrongs ascribed to them. Some more specific information will be given in the subsequent chapters in connection with these two diseases, but before going on to the subject of shepherding, it is well to try to get the shepherd to recognise that these diseases exist, and that the want of this recognition has been misleading to him in many ways.

It is very commonly held that the sheep is a delicate animal, always ready to take ailments, and, having taken them, being of such poor constitution that it will, in most cases, die. In fact, the old idea that it is of little use to doctor a sheep still holds, though not to such an undeviating extent as in the past. Sheep are not really weak-constitutioned animals; but because it is not until sheep are seriously ill, or inconvenienced, that they show symptoms of distress, it has been assumed that they are delicate. Sheep in domestication lead widely different lives compared with their forbears who lived in a wild and indigenous state. One may say that practically everything is altered; they live according to the will of man, and not to Nature. They are taken to conditions altogether different; they are fed with foods quite unnatural to them, and often they are kept within a very narrow range, enforcing them to cover the same ground with quick repetition, and they are maintained in large flocks. Moreover, the "improvement" wrought on sheep to increase their flesh-forming capacities, and the yield of wool, must have some constitutional influences foreign to them in a natural condition, and not necessarily to the maintenance of their powers to resist hardships. Unlike the horse, ox, and pig in domestication, they live their whole lives on the land, and are uninterruptedly dispensing such internal and external parasites, and some one or other of germs of specific diseases as may attack them, on to the land. Therefore the land becomes so infested with them that other sheep following them can scarcely fail to contract the same ailments from which they suffered. That is a common experience in over-populating, whether it be with man or any of the lower animals. It is only by keeping fewer animals, and resting the land for longer periods, so that the parasite or germ may be starved out, or by following such controlling methods as experience or science has pointed out as effecting disease eradication, that reasonable hope can be entertained of keeping the sheep in health. Successful shepherding depends very much on the thoroughness with which diseases and ailments are kept in check.

Without being able to ascribe a specific reason for their actions, farmers and shepherds have recognised that land becomes unhealthy if overstocked, and that conditions indicate that they must rest the land. They were not to blame, because, until the man with the microscope came along and saw what no man could see with the naked eye, they were forced to remain in ignorance of certain diseases. No one knew they were there, consequently no one could tell how or when they were contracted. No wonder, then, that when sheep ailed, and for no particularly apparent cause scoured, withered and died, in spite of all that sheep-lore could suggest, sheep were regarded as being delicate. Moreover, it is ever the shepherd's duty to try to discover the cause of ailing, and to supply the remedy, and to his credit it can be said that, where the cause is not deeply submerged, he very frequently arrives at the right conclusion. But when the cause is due to something that no man has divined, he is bound to be at a loss to attack it in the right place. Following his usual and right practice, he has to look for something more obvious, and when one knows that these until recently hidden diseases existed, his "shots" at the cause seem to be ridiculous. As a matter of fact, they show only the more conclusively how necessary it is that all encouragement should be given to those who systematically endeavour to trace out the nature, origin, and prevention of the diseases which affect sheep.

Not many years ago a prejudice existed against the use of artificial manures, because it was held that they produced unhealthy food for sheep—the sheep went wrong when consuming crops manured with them. It is very easy to understand how shepherds with flocks troubled by a mysterious disease should jump to the conclusion that it was due to something immediately influencing them ; no other cause presenting itself, what more obvious than that the manures, which they did not understand, should be the cause? And very generally the manures had to take the blame. I have heard specific charges brought against artificial manures for rendering almost every kind of fodder and forage crop injurious to sheep by men who, with several further years' experience, would not for a moment accuse them of it now.

Probably the most astounding misconception, and consequent mistreatment, of a malady in modern sheep farming is that which attaches to the very fatal infestment of the fourth stomach of lambs, and sometimes sheep, by minute worms of the strongyle type. In parts of the North of England and Scotland a practice has grown up (and is probably of considerable antiquity), in districts where losses from these worms are frequent and heavy, of operating on the newly-born lambs in the following manner to prevent or cure the attack : the bony protuberances over the eyes of the lambs are bashed by a hammer or stone. That is all ! No one seems to understand why it is done—but it is. Yet how the bashing in of bones in the head is likely to prevent scour and wasting away through derangement of the digestive organs requires an imagination which it is difficult to conceive. It may be that it started through someone having heard that by bashing in the soft, bony swellings above the hydatid that grows in the brain of sheep, causing gid, thus killing the worm and sometimes curing the sheep, thinking that by breaking any protuberance on the skull any other disease might be cured ; or he may have confused the whole matter. However, when such things are possible, nothing seems impossible ; and, as a matter of fact, some very extraordinary practices do exist in shepherding, though the more frequent calling in of modern-trained veterinary surgeons has done much good. With all the skill of shepherds, and their possession of much valuable sheep-lore, it cannot be denied that there are many practices requiring eradication ; farmers should therefore equip themselves as thoroughly as possible with a knowledge of veterinary science, so that they may be able to discern the good and the bad in the methods adopted by their shepherds. Much has been handed down to shepherds in the form of sacred secrets, of which they are very proud, but which the sheep thrive in spite of— some are not so harmless. With it all, if a veterinary surgeon proves that his methods are better than those which have been previously employed, experience shows that shepherds are not too prejudiced to recognise it, but it must be very clearly demonstrated to some before their pride in their art will allow them to admit it. The general use of antiseptics and more hygienic management in the lambing-pen are illustrations of the readiness of the shepherd to change his methods when he is convinced that by so doing he will get better results. What is truly necessary is to provide means by which shepherds may be kept conversant with sound, new practices.

The above remarks may suggest more than is intended in respect to the shepherd's want of technical knowledge. The intention is not to disparage his skill and value, but merely to indicate that there is fresh knowledge to be gained as investigators make discoveries, and the desirability of having this knowledge put before him as it becomes available. The responsibility of doing this really falls upon the farmer, who has better opportunities of acquiring knowledge of the origin of diseases than has the man who spends the greater part of the day in hard work among sheep.

BRITISH SHEEP AND SHEPHERDING.

CHAPTER I.

INDIGENOUS BREEDS.

The wealth of meat, the splendid fleeces, and the early maturity of the modern breeds of sheep are familiar to all, and it is only fair to the great breeders who have built up these breeds that those who reap benefit from their work should appreciate what has been done for them. There is another aspect, and that is, that it is well that the indigenous features which were originally involved in the making of a breed should not be lost sight of. It has to be remembered that up to the present time there have been men who could remember sufficiently far back to recall many breeds which were only just emerging from the crude, unimproved state ; more than that, there are old men who met men whose recollection dated back to the time when the earliest efforts in the modern phase of sheep improvement were in progress—men who met Bakewell and Ellman. These, however, are few, and in the course of Nature they must soon pass away. With the older improved breeds, direct observation and hearsay can no longer be relied upon to keep in remembrance the features of the sheep in an unimproved state, and all that can really demonstrate the features of these animals are those illustrations which truthfully depicted them in those early days. Unfortunately, very few of the portraits were faithful—for they were not drawn or painted with technical accuracy. Fortunately, there is one excellent source of information in the series of paintings unearthed by Professor Wallace at the Edinburgh University ; these were the original paintings from which the coloured plates accompanying Low's " Breeds of the Domestic Animals," published in 1841, were taken. The originals were painted by Shiels, R.S.A., and the drawings for the plates were made by Nicholson, R.S.A. The object of the set was to reproduce the breed features and characteristics, with colour and relative size, so as to make a faithful record, rather than a mere picture. Nothing before or since in animal portraiture that I have met with can compare with these in an educational sense.

Coleman's "British Breeds of Sheep" was illustrated well by Harrison Weir, and, published in 1877, is valuable as showing the breeds about midway between Low's time and the present; but Low wrote when there was only one Down breed, and when the Hill breeds had been subjected to nothing more than desultory crossing; the Down breeds of to-day were in the melting-pot, and the moulding to present lines had only been crudely outlined. What the Down breeders have done since that time is to make a record in breed-making that stands out beyond anything ever accomplished with live stock. Of course, before the several breeds, not only of Downs, but the whole range of those which under improvement have come down to us, took definite form, much had been done by the geniuses who handled them in the crude form and, by judicious mating and selection, built up the types. They were the men who had to do the spade-work that made other things possible. What they had to work on is shown in the published illustrations. What they did, however, in no way belittles the great work done by breeders since, and never more actively or with greater success than at the present time. If in subsequent pages, when dealing with the several breeds, I have not dwelt as enthusiastically on the merits of the individual breeds as an enthusiast in a particular breed might be disposed to think they warrant, it is not because of want of merit, but because all are so full of merit that it is unnecessary to impress the fact by redundant repetition. No one appreciates better than I do how grand our British breeds are, one and all. If some remarks are critical, they express merely my personal view, rather as to where things may lead to, than as indicating present deficiency. But I believe we are in some respects at the crossing of the ways in several aspects of sheep-breeding, and it behoves breeders to be careful of the weapons that they use.

If any evidence were needed as to the probability of some changes, one has only to look at the vast amount of cross-breeding carried on over large districts where little was done until comparatively recent times, and to the modern tendency to adopt breeds in accordance with their usefulness in their district, and not to rely merely on the local choice, which was established when conditions were probably very different from those which prevail to-day. There is certainly more intelligence in the selection of sheep to suit local conditions than was met with in many instances in bygone days; it is being met both in respect to pure-bred and cross-bred sheep. In other pages these points are referred to from time to time, so more need not be said here. But the indigenous features of the breed bear much on this point, and it is for this reason that it is desirable to examine the breeds as they were when the indigenous features of the races which entered into their composition were very apparent.

As illustrating in what direction these observations might be made, there are shown the old Wiltshire breed, the Southdown, the Cotswold, and Leicester, which, with the Berkshire Knot, or Nott, in varying degree formed the basis of the Hampshire Down, the two first and the last playing the main part. The Wiltshire was the only horned breed, but it took a long time to eradicate the horns entirely, the recurrence of snags up to within comparatively recent years being evidence of this. But the pronounced Roman nose came from the Wiltshire, as did the long, drooping ears ; their presence is valuable as denoting the fact that the indigenous blood is there, making the breed adapt itself to the district in which it is associated. The dark face, mainly from the Berks Knot, is evidence of the retention of that blood, and therefore a suitability to heathland other than the chalk heaths. The Cotswold helped in size, and also fits a breed which, if only the product of dry, chalky soils in a dry climate, would not fit it to winter on cold, wet arable land. The Leicester, although the evidence of a liberal introduction of this is small, doubtless has had influence on the wool and clearness of the skin, as well as of the powers of early maturity. The Southdown gave the Down characteristics, with the thickness of flesh, fine quality of the fat, and the denseness of the fleece. With all these characteristics, it is not surprising that the building up of a breed essentially to suit the conditions of downland should be based on a constitution that adapts it to many other soils and especially to go very wide to cross with other breeds, notably with the longwools. The fact is it does. Hampshire rams are greatly in demand, and increasingly so for this purpose.

The reproduction of the engraving of the fattened Berkshire Knot wether is the only portrait I have met with, and was published in 1803, when already the breed was being much crossed out by the Southdown. It is typical of the hornless local heath breeds found in various parts of the country, although the heath breeds were more commonly horned. In making the composite breed out of these several stocks, breeders had to find the proper proportions to instil, and to maintain or eliminate the features which they approved or disliked, and in these they were naturally guided by observations which they made in the field and fold. That they did it well there can be no doubt, and when they commenced that system of in-breeding which is part of the fixing of type where crude and irregularly bred animals are brought under improvement, it is easy to imagine how some improvers gave greater prominence to certain features than did others, and how healthy rivalry sprang up, just as it does to-day; but then they were much more face to face with raw indigenous characteristics than they are now.

Having fixed a type, and produced an animal that could greatly influence for the better those which had no fixed type, it is easy to recognise how farmers welcomed the flock-improving rams. Incidentally, it seems to be very often forgotten by those who see virtue in narrowly made breeds, and claim that their breeds were never crossed, that the rams from the pioneer breeders did not go on to pure breeds, but on to the ordinary stock of the country, and that this ordinary stock was very variously bred ; also that many of our best flocks have come from flocks thus established, and through which, often in very thin streams, no doubt, but fortunately still there, the blood of a mixed ancestry runs. As a matter of fact, there is practically no really single-blood breed, as far as can be shown. If our breeds had been of single blood, they would never have possessed their great cosmopolitan powers of adaptation. From the coarse, horned, whitefaced old Wiltshire to the modern Hampshire is indeed a triumph for the breeder.

As a more modern development, and one accomplished to a large extent by those living, may be instanced the Suffolk Down, which had as an indigenous foundation the long-legged, horned, unthrifty Norfolk horned sheep, and as its improver the Southdown.

The Cheviot ewe shown represents the little hill breed after it had been influenced by the Lincoln. Although having been shorn, she shows the characteristic Cheviot frill that is left behind on any breed into which it is instilled, and is so marked in the Exmoor, Romney Marsh, and Ryeland—three breeds which, under the influence of the help given long ago, are coming so fully into the public eye at this time.

The Leicester, of all the breeds at the time when the paintings were made, shows the type and characteristics associated with modern sheep; it stands among the other breeds—with the exception of the Southdown, perhaps—as the only modern breed, and it is little wonder that it was used so universally throughout the country, alongside of the Southdown, to convert the unimproved breeds into thrifty sheep. The toning down of the old marsh-bred Lincoln, enabling it in our day to become so appreciated in all wool-growing countries, is an evidence of its influence, though it was shown on the other longwool breeds throughout the country, and was the forerunner of, and great instrument in, the developing of early maturity in British breeds of sheep.

With these instances the other breeds can be examined, and with the knowledge of the origin of any breed, it is easy to recognise why certain features have been kept prominently forward in them, and how they have been modified to suit changed methods of farming, and the requirements of the meat and wool markets. This accounts for many so-called " showyard fancies," which, after all, are essential features retained as hall-marks to demonstrate the blood that was used in the making of any specified race. As has just been said, there is a risk of these indigenous features being under-estimated or over-estimated, because there are no longer those who remember the original stock in its indigenous condition, and it is easy to miscalculate their value ; consequently, enthusiasts on some point, unless very careful, may run riot on the point of fancy. In the history of our breeds a stage has been reached when this should be guarded against.

CHAPTER II.

The Influence of the Breed-maker.

The breeds of sheep may be classified in several ways : (1) By their features as influenced by their historic origin ; (2) by wool characteristics ; (3) by nature of the land, its richness, altitude, &c., to which they are indigenous ; (4) by their mutton-producing properties ; and (5) by the intermixture of breeds which have been incorporated by the modern breeder. However, these to some considerable extent merge at points, and a more general division is most practicable. For practical purposes the division into (1) sheep of the rich grasslands (2) sheep of the arable land, and (3) sheep of the hills, heath, and other poor land, is as serviceable as any. Low, who some seventy or eighty years ago wrote so ably on sheep in that delightful book, " The Domesticated Animals of the British Islands," made a distinction between the sheep of the higher hills or mountains, the sheep of the heaths, and the sheep of the rich lowlands, and it has much to commend it ; but in his time there was no Down breed recognised but the Southdown. Since then several breeds originating from heath breeds crossed with the Southdown have been recognised as established types, such as the Hampshire, Suffolk, Shropshire, &c., all of which carry many features of the Southdown—in fact, they are called Down races. But the Southdown itself is a heath breed, the Southdown Hills themselves being somewhat rich chalk heaths. By accepting the term heath as representing those sheep which originated on more or less indifferent pasturage, otherwise than on that found on rich lowland, or the higher hills carrying poor feed, a good many points of confusion in respect to the indigenous types of sheep are removed. The difficulty in respect to the term mountain as applied to sheep is that some of the sheep commonly found feeding at the higher altitudes of the relatively very low mountains of these islands (in comparison with the mountains of many other countries) have common origin with others that have never been associated with the heights. There are, however, some which show no recent (though it be several hundred years', or possibly thousands) direct relationship with the sheep of the heaths or richer plains. Although the evidence is difficult to confirm definitely, there is reason to believe that, in the distan past, sheep, like man, came westward from Asia during great migrations—the former probably accompanying the other. The short-tailed, goat-like sheep still found in the extreme north, and in the hills of Kerry, Ireland (though in recent years they have been nearly crossed out), has its counterpart in the north of Europe and north of Asia. The many-horned sheep still found on some islands off the coast of Scotland are related to the old Iceland breed.

The soft-woolled sheep of Wales, the white or pink nosed, somewhat antelope-shaped sheep of the type most commonly met with in Wales, is a long-tailed variety, of a tan-faced type not extinct in Scotland in Low's time, and thus differed from the short-tailed northern breed—in fact, having more in common with the sheep of the Celtic nations of Europe; moreover, they are held to be of even more remote antiquity than the short-tailed. One other true mountain sheep is the goat-like breed of the higher Welsh mountain, though with tail of natural length. It is a true mountain grazer, and prefers the highest altitudes, though if brought lower it increases in size and loses much of its wildness, and Low believed it to be the only descendant of the ancient sheep in South Britain. These are what would be regarded as the more truly mountainous breeds in this country. In fact, it is doubtful if other breeds introduced later, as by the Romans, no matter how they have altered in the course of time, are strictly entitled to be regarded as mountain breeds, unless it be one or two minor ones, the origin of which in this country is traditional. These old types have intermingled very little with the better breeds.

Leaving out the ancient breeds which have their homes in the high hills and more remote parts of this island, we come to those which are of greater economic importance. The heath breeds and the breeds of the rich plains, in the main, represent the great divisions, the long-woolled and the short-woolled breeds, and the less often used, though desirable addition, the medium woolled. The longwools are more essentially grass-raised on rich land; the Down-developed heath breeds are suitable to pass much of their life on the arable land; whilst the heath breeds which have not been influenced by the Down thrive under more trying conditions generally on grass land giving only moderately good herbage, and are not kept much on arable land in their native districts. Considerable licence may be given to this division, but no classification can be made that does not break down somewhere; it is given as a general survey.

Although there are many recognised pure breeds of sheep, generally taking their names from the county or locality from which they sprang, very few, if any, can claim to having sprung from one stock, or from stocks from one district only. Anything like systematic sheep improvement was unknown until Bakewell started to improve the Leicester, and he carried out no experiments before 1755; consequently nothing was done before 160 years ago. Much has been done since. The position up to this time was that there was relatively a small area of land under the plough, as compared with that a century after, tillage poor, manure deficient—in fact, it was much as it had been three or four centuries previously. Going further back, the greater part of the country had been under forest and wastes; much land was unenclosed; sheep were wool-producers and manure-carters, carrying their dung on to the fallows at night after feeding on the grass by day; there were no turnips; no clovers were grown in rotation; in fact, when not folded they roamed at will in many instances. Where there was little arable land they were rarely folded. Having nothing but what was indigenous to the land, their constitutions adapted themselves to their food and environment; and over districts as far as the food, soil and climate were closely the same, the sheep took certain characteristic features, such as we should now call breed characteristics. In some instances great similarity existed over wide areas; in others, in very limited ones; in fact, these breeds or variations of breeds were almost innumerable, although in many instances the origin was a common one in the remote past.

There was one very noteworthy feature: the sheep, at any rate where they existed in considerable numbers, which had developed on a certain type of land, food, climate, altitude, &c., recognised in one of those instinctive ways difficult to explain that if they were to be mated with sheep developed under other conditions, the offspring would suffer; and though sheep produced under two totally different sets of circumstances met at the point where the circumstances differed, they would not cross-breed. Nor would strange flocks brought together mate, unless intentionally confined to do so; each would separate and graze in different parts. Moreover, this instinct remains, although many generations of man-controlled breeding have been brought to bear on them. Had there been indiscriminate inter-breeding in the old days, there could have been no pure strains or breeds; all would have been mongrels, as were those which were kept in small numbers and bred as they met.

The good work done by Bakewell on the Leicester was soon shown, for his improved sheep had great influence on other breeds with which they were used. Then Ellman, of Glynde, who had started in sheep improvement some few years subsequently to Bakewell, greatly improved the Southdown. With one or other or both of these breeds, sheep improvement throughout the country made vast strides. For a time there was no particular method in a general way; but gradually other skilled breeders arose, and these began to select the best of their breeds after they had been brought under the influence of more improved animals. The extent to which what were known as Improved Leicesters and Southdowns were used on the breeds throughout the country is realised by few to-day. The influence was in most cases useful, but when breeds which were not suitable were mated, and although the sheep might show outward improvement, harm was occasionally done by producing an animal not so well fitted to live the life the conditions of farming demanded as were the original stock. But there was practically not a breed that did not benefit from the cross with either the Leicester or the Southdown. At times the cross was too violent, but in many instances to-day the effect of those early crosses is noticeable, to the advantage of the breed.

From the crosses originated at the end of the eighteenth and beginning of the nineteenth century most of the notable breeds of to-day sprang. Some came earlier than others, and some have been established as pure breeds (though admittedly composite); whilst some of our best breeds are the produce of two or more composite pure breeds; and the work of creating new breeds is going on—and going on with much vigour—in fact, one looks with considerable confidence to seeing much breed-making among sheep in the next quarter of a century.

The great breed-makers of the past century or so conceived one fundamental idea—that certain indigenous features should be preserved. As it is difficult for those who have seen only the highly improved races of to-day to conceive from what ill-shaped, unthrifty animals they were developed, some illustrations of the breeds as they existed a century or less ago are given here. Few inventors or discoverers in any industry have done better work than have the men who worked on the unimproved live stock of this country, and developed the breeds of to-day. Even with the knowledge of what has been done there are not many who could take up one of the few remaining unimproved sheep types such as Mr. Elwes has brought together, and so select and mate that the offspring in a few generations would be materially improved. It is easy to cross them with improved animals, and get striking results, but the position is very different to that when there were no improved breeds, and a mind-picture had to be made as to what should be aimed at. How well these pioneers did their work; with what foresight they pictured the improved animals; how they eliminated features that were undesirable; and how they built up a science of breeding where no science had been; must always place these breeders among the most valuable and most accomplished of those who have shared in the development of modern agriculture, or this country's industrial advance.

Nor should the excellent work of those who have pioneered more recently go unrecognised. Breeders have hitherto been careful to retain some of the indigenous features of the breed that they have undertaken to improve ; and to keep in some characteristics of such other breed or breeds as they have used on it ; and they have blended these so nicely that they are beautifully balanced ; no feature seeming to jar on the eye, or to destroy the useful purpose of the sheep.

But a rather dangerous period has been entered upon, and breeders have need to be very careful in their methods. The welfare of the sheep in the future is in the hands of the breeder of to-day. Just as in the early days of a breed's improvement " fancy," that is, the fancying or thinking that a certain feature should be retained or eliminated, was the subject of controversy among those who were engaged in the remodelling of type, so fancy continues in all stages of development ; though with so much to guide them as there is now, in comparison with that available to the early pioneers, the opportunities occur more rarely. There are always some, no matter what the breed, or for that matter, animal, who are more impressed with the value of a certain feature than are others, and this is often shown in the show-ring, where one judge will upset a previous decision. One man has found that on his land or for his purposes one feature or set of features give him the results he wants and at which he aims ; another equally experienced man finds that other points are more profitable to him—each may be equally right according to his experience. But a third may have a strong fancy for something which is not so valuable ; and he may be a man who has great influence over others, and his less correct opinion may be accepted and acted upon to the prejudice of the breed.

It is most highly important when a man pioneers in a breed, even though the breed may have been brought to a high point of perfection, that he should know why those before who did so much for the race retained or specially developed each breed feature. There are certain features common to all breeds where the early maturity and constitution of the animal are concerned, and to which all must conform, but breed features are a thing of themselves, and it is highly important that the mistake is not made of trying to over-develop some feature suitable to some other type, and not so valuable—possibly objectionable—in the breed one is handling. There is so much evidence of the skill of our breed improvers in being able to develop any feature they desire that no one doubts their ability to accomplish whatever they set their minds to ; though no one has yet accomplished the act of making a breed possessing the quality of the Southdown or Suffolk which will carry a fleece of the weight of a Lincoln. It may not be impossible, but it is far from being realised.

Great weight of wool of finest quality, with a carcase of higher class mutton—that is with a great preponderance of lean meat of choicest quality, and with but little fat when matured—is the natural aim of all breeders, but there is great natural antagonism to its realisation. Good wool and good meat are found together, and when the weights of wool which many of the breeds now among our best produce, are considered with the very poor yields which were given when the ancestors of these breeds were in an unimproved condition, it has to be admitted that great work has been done. But great weights of fine wool are obtained from breeds which have a special tendency to develop fat on the backs ; in other words, heavy fleeces grow out of big beds of fat, and the fat-backed sheep—the longwools—produce mutton that does not rank as first-class. Moreover, those who can afford to pay a high price per pound of mutton will eat very little fat, and it is only in a few districts where a superabundance of fat is appreciated even among those who cannot afford the highest prices.

In view of the fact that the English sheep raiser is preferentially placed in being able to market his mutton in a fresh condition whilst the foreigner has to chill or otherwise preserve his, it is obvious that whenever circumstances permit it is desirable to aim at producing mutton of the highest quality so as to secure the highest prices. The natural conclusion to draw from this is that the quality of the meat should not be sacrificed to the yield of wool. In practice farmers have appreciated this, because for a number of years farmers where the indigenous breeds have been longwools giving poor quality meat have seen the advantage of crossing their sheep with Downs to improve their meat at some sacrifice of the wool. This in itself is a good practical proof of the farmers' view of the matter—they want better mutton, and they can make more by improving their mutton than by maintaining the fleece at its old weight. If it were not so, crossing longwools with Downs would not be followed to the extent that it is. At the time of writing, all mutton is easy of disposal ; but in times of over-supply or of trade depression, mutton of poor quality is very difficult to sell at remunerative prices, whereas high-class mutton is always saleable.

This gives rise to the question—are modern methods of pure breeding not being carried out with, or having a tendency towards, too much observation of the wool interests at the expense of the quality of the meat ? This is, admittedly, a very controversial question, but it is asked with the purpose of producing an illustration showing where a good feature may be developed to excess. Other features might be dealt with, but the one selected is one which many breeders have set great store by, and it is reasonable to argue that its over-development may have a bad influence; although moderately developed, it may in many instances have a good effect. It is that of the overcovering of the head and face with wool. In speaking of the bad influence that is meant to relate to the British mutton producer, though it may be specially attractive in those countries where the value of the fleece has a different relative value to the carcase. There are certain minor points favourable to the covering of the poll of sheep. Some few breeds with thin skin on the poll are liable to irritation by flies, and wool on top is protective. A well-covered head gives an attractive sweep from the nose over the head, suggesting a better outline to the fore end; it also suggests a full fleece.

Head-covering becomes objectionable only when it is overdone. At the same time it cannot be held that the Suffolk sheep, which is bare from the setting on of the ears, loses anything in style or appearance because it is not covered. As a matter of fact it is suggestive of the high quality meat for which it is so properly reputed. Whilst the Suffolk sheep holds its high position for meat it cannot be urged that head-covering is essential; and in respect to that breed, it is not required for protection against flies. On the whole, rather more than is necessary has been made of the point of protection from flies, except in the case of a few breeds. There are many things invidious and unfair in comparing breeds, because certain features are retained in each one to ensure its continuing to be properly suited to the soil and climate of the district to which the foundation stock was indigenous. For this reason there must be distinctiveness; it would certainly be a grave misfortune if all breeds were merged into one, if it were possible, which is not a practicable proposition. This country holds a strong position because of its varied types of sheep, with their different aptitudes and characteristics; and it is essential that those who control and pioneer in breed modifications shall be quite clear as to the wisdom of the direction in which they move.

During the past twenty-five years whilst head-covering has been so much in vogue, it can scarcely be said that the breeds which have responded most to the endeavours of breeders in many respects have made a corresponding improvement in the quality of their mutton; at the same time there has been improvement in respect to earlier maturity, and in skin and fleece. But it can be said that the Suffolk, a bare-headed breed, has made great strides during that time in both mutton and fleece. This is not a surprising result. It is not unreasonable that some advantages may result when a harsh, unthrifty breed with little aptitude to fatten is taken in hand, to encourage head-covering, because such a breed has to undergo radical change, and this implies some necessity for increasing the fattening propensities. That face-covering is not a necessity has been proved by the Suffolk in its translation from the old Norfolk sheep, which was a typically unthrifty breed when taken in hand. But breeders who saw that they were securing some advantages in other directions, whilst they were getting more head-covering, seem to have put too much value on the head-covering, and some have taken it too far. They have developed, especially among some of the Down breeds, too great a tendency to produce fat, and the quality of the meat has suffered. When they were producing sheep well covered, they had to go to sheep with a special tendency to wool production; thin fleeced sheep would not have that exuberant tendency to wool production that was necessary to make its growth spread on to parts which for centuries, and possibly from all time, had not carried wool.

But sheep with great wool tendency are fat producers; therefore, in selecting their sheep from the sheep possessing the greatest fat and wool tendency, they have developed sheep which too readily run to fat, and the proportion of lean to fat has suffered. They have tried too much to imitate that essentially wool producing sheep—the Merino. Though it is shown in several of our long-wool breeds that head covering is not essential to the production of a heavy fleece, it is probable that the efforts to introduce the Merino cross a century or so ago, commonly regarded as a failure, have left more influence behind them than is generally recognised. The value of the wool growing on the nose and below the hocks of sheep, is little more than the expense of cutting it, and in the case of sheep much kept on arable land, excessive wool below the hock is not desirable, because of the tendency the wool has to collect dirt. But a good fleece on the belly is valuable, as sheep provided with it lie warmer on cold ground. It is satisfactory that the Shropshire Down breeders who were early sinners in respect to the development of wool on the nose and down to the fetlock, have sufficiently recognised the position to hold that in future *excessive* head-covering will not be regarded as a point of merit.

It does not seem to be recognised sufficiently that our British breeds are not genuine pure breeds; but that many of them are derived from deliberate crossings, involving two, or as many as five or six, different breeds; and also that apart from this there was that casual crossing when the Leicester and the Southdown, more than a century ago, were first put on the native breeds. Nor is it recognised that it is not at all necessary for that blood to have been bred out even by this time; there is no particular reason why it should be. Where flocks have been kept with careful breeding for a long time in straight line, it could not go out, although it may be attenuated.   There are certain characteristics left behind, possibly very difficult to demonstrate, but so long as they are there, they can be encouraged through selection to come into greater prominence.

It is held that very few sheep escaped crossing with the Leicester during the first forty years after Bakewell improved that breed. All this was shown before there were any breed societies or controversies as to breeds, and doubtless many who have held that their flocks have never had any outside blood in them, but are truly indigenous, have not known all.   The Cheviot, too, did much useful, though less-recognised, work.   When selecting to head-covering, involving the selection of those with a fat and wool tendency, it is pretty certain that, as the old Heath breeds which had little aptitude to produce fat, could supply these only in a very moderate degree, they had to come out of the more fatty breeds which intermingled with them at a date previously to the time when, mainly during the last century, the new breeds were evolving.   Selection to this excessive head-covering, with its accompaniment of the tendency for producing more fat, has really meant the bringing into greater prominence the fat-developing features of a remote cross, probably most often of the Leicester, but at any rate of one of the fat-backed breeds.

If it is worth while to maintain a very large portion of the British sheep as producers of mutton of highest quality, anything in the way of selection that works against it should be guarded against. Much space has been give to the argument of this point because there is unfortunately a tendency on the part of those doing excellent work, otherwise in bringing to the front breeds that have hitherto been regarded as being of minor importance, to repeat the error which has been committed on some of the more prominent breeds.   Before they go further they had better learn what the Suffolk men have done with their breed, and thoroughly digest it. The popular fancy at the moment is for a covered head.   The foretop of the Oxford is being stamped out of recognition, though possibly not more than there is a call for in some directions.   But the Oxford always had a tendency to fat through its more direct association with the Cotswold.   It suits districts where some improvement in the meat is required, without the loss of size, and there is no doubt that an infusion of Cotswold in a Down breed enables the offspring to thrive on arable land on the oölites and other cold soils, as they cannot thrive to the best without it. Consequently there are districts where the Oxford is a more popular cross, though over a far larger area the Hampshire is preferred. But if the Oxford men " Hamp " their breed too much they will, in course of time, lose the distinctiveness, because the difference is one of degree only.   The fact that so distinctive a forelock as the

Oxfords has been so much modified shows how great is the worship of this fetish—the wool-smothered face.

The remarks which have been made bear on the face-covering as it affects British breeds which are specially favoured in having, whilst they mature without excessive fat, the advantage of supplying the market with the highest-priced mutton—the home, fresh-killed, unchilled, finely-flavoured and finest-grained mutton. The foreigner can find his way into any of the lower grade markets, and it is surely worth while for the home breeder to cater for the unassailed markets. But British breeders of the races which have found most favour with the wealthy foreigners who have loose purse-strings, cannot be blamed for having paid so much attention to the wool aspect ; and up to a certain stage it was an advantage to the British flock keeper to have the sheep improved both in wool and meat ; they could afford to go alongside for a long distance ; but now mutton stands out as so much the more valuable portion of the sheep in the breeds of highest meat quality, that the greater interest should be most regarded. Up to the time before chilled meat became an important factor in our markets, the foreigner in many countries regarded the sheep almost entirely as a wool producer, and in searching British markets for improved stock his first consideration was the wool. It was obviously the soundest policy for the leading breeders who scooped up most of the big prices paid by the foreigner, to cater for him. One cannot feel otherwise than that in doing this the British interest was not always the first consideration. Let it be understood it has not suffered profoundly yet ; and in this lies the greatest hope for the future. There is time to remedy what has gone a little to excess ; but it is time to grapple with the question ; the danger lies in the fact that minor breeds which have such a grand chance to come into prominence and be of greatest value to the mutton producer, appear to be following headlong into the error which has gone before.

It is time for the supporters of major breeds carefully to observe the lines on which they are travelling, and great care should be taken to model the new breeds on the soundest lines. The material is there ; it is all a question as to how it will be handled. It must be remembered that the home interests are not identical with the foreign ; they can run parallel for a distance, but there is a parting place, although mutton stands in a very different position in regard to wool in the estimation of the foreign grower than it did before he could ship his mutton to all parts of the world at a remunerative price. The harm done is not serious. What has been put forward has not been said to imply it ; it has been given rather in a cautionary sense to warn against overdoing in the future. The sheep stock of the world at present is short, and liberties can be taken now which would be far more serious when the world's stock gets in excess of the demand.

The suggestion that changes in the distribution of sheep and the making or remodelling of breeds is not prophetic, as it is actually in progress.

The leading supporters of some of the most renowned breeds admit that they have been brought to a standstill, and can see no prospect of further advancement on the lines they have followed; at the same time, they recognise that other breeds which had been in an insignificant position and had not challenged supremacy with them until recent years, are now strong competitors. These men are not blind to facts, and recognise that they must take drastic measures to hold their position, although it be by recasting their models by bringing in new blood. There is nothing contrary in this to what has been done before. Our breeds, almost without exception—and these generally insignificant breeds—have been built up by the amalgamation of two or more types. Those who have watched the breeds for a few decades cannot recognise other than that at times some of them have been recast by help from breeds not popularly supposed to be included in the accepted composition of the breeds. Rather more than twenty years ago I mentioned in an article on Cross-breeding in Sheep, published in the Journal of the Royal Agricultural Society, when treating of pure breeding, that "even in some high-class flocks it is often difficult to believe that some stranger has not been there."

The late Mr. Charles Howard, of Biddenham, than whom no one stood, in his day, in more repute as a breeder and judge (in fact he was referee for all classes of stock at the Jubilee Royal Show at Windsor), in discussing the article with me at the time, said that there was no doubt as to its being done. Moreover, he implied that he recognised it in the breed with which he was most intimately associated. The fact is, changes in type have sometimes been too sudden for them to have been brought about by ordinary selective means. Probably much good has been done in this way; but when it is done, it should be notified to the breed society most affected, and the society should not allow animals so influenced to enter the show ring until it is assured that all risk of the injuries from cross breeding which mongrelism can give rise to are removed. The change and improvement in sheep breeding have always gone hand in hand with improving farm methods, and these changes have been essential. The future may have some additional conditions in farming to make it necessary to alter the types of sheep to make them adapt better to them; but the likelihood of this is difficult to see, at any rate in any marked degree. It therefore rests very much with the future to deal with the material at hand, and to produce such sheep as best meet market requirements. This view has already been gripped by farmers, and considerable changes have taken place in recent years. Whether the land will go into far more general small-holding occupation is not assured yet. Legislation may do all it can, and men may be tempted to acquire small holdings, but whether they will be induced to remain there is quite another question, which will not be solved until agriculture in the ordinary evolutions from prosperity to depression which come from time to time, meets with a period of bad times. If they stand through a long period of bad times they are likely to be permanent, but the past has revealed much that suggests that many would flee from the land.

In a small-holding community sheep have a small place; the future of sheep farming hangs much on what occurs then. It is best to assume that the conditions of farming will not greatly alter, and to proceed with sheep improvement as though it will not. It will be a poor day in British agriculture when conditions compel a large diminution in the head of sheep carried. Sheep have made and marked its progress, and still are mainly responsible for its steady success.

Where the change in sheep farming, in respect to the distribution of breeds has been most shown is in the less servile adherence to preconceived ideas. Far too much stress has been laid on the want of wisdom in trying the usefulness of other breeds than those hitherto kept in a district. That injudicious introductions have been made through not paying attention to conditions and circumstances, cannot be denied, but too much capital has been made out of this by those who have wished to get credit for sage counsel. To look before you leap, and to make trial before you embark heavily in a strange breed are equally sound advice; but one is allowed to look and to make trial; though many, very many, have feared to make trial because the unwisdom of change has been so generally taught. In the more businesslike enterprise of to-day, however, such advice does not go unchallenged, and breeds foreign to many districts are ousting those which have long been regarded as permanent fixtures. Breeds are moved to new districts to be kept as pure types or for crossing. English sheep go to Scotland, and Scotch sheep come to England. The black-faced mountain breed of Scotland returns to the lowlands from which it migrated as a heath breed long ago. Outside some districts it is little recognised how great is the migration of Scotch sheep to permanent occupation in England. Probably, however, the greatest influence on English sheep breeding—it had a big one in Scotland before—is the Cheviot. The Cheviot seems likely —within the course of a few years—to have as great an influence on English breeds as the Southdown has had. If ever a breed needed to be well guarded, so that its best usefulness may be secured to other breeds, it is the Cheviot at the present time. It cannot fail to do much, but what is essential is that its characteristics as a heath or hill sheep shall be maintained, so that the quality of its mutton shall not deteriorate. This is mentioned because there is some fear that those who have first interest in it may try to develop it too much to the type of the Downs, with their wool-covered faces and increased tendency to fat at the expense of the lean meat. This applies also to two other breeds which have undergone great improvement in recent years—the Ryeland and the Exmoor.

CHAPTER III.

BREEDS AND THEIR MODIFICATIONS.

When Bakewell and Ellman commenced their work, it has been previously said, that many breeds, with almost innumerable offshoots, had formed, and were met with in their indigenous homes in all districts. Many of the best of these came under the influence of men of special skill in breed-making, and they, with variations, are reproduced in the breeds of to-day; but there were others which lacked such men to guard their interests, and they, to a great extent, disappeared before the march of improvement in sheep-breeding. The partly improved sheep were found more profitable, so the indigenous stock either disappeared or was merged into some strong breed near by. When the indigenous race was wiped out it could not be replaced; the consequence is that there are many areas in the country that carry stock that is alien to it. It is these districts which, even yet, often need an imported breed indigenous to conditions closely resembling those which obtained before the old breed was wiped out. It is obvious that sheep bred exclusively on light chalk land for centuries would not have the same type, constitution, or character of those bred not many miles away, but on totally different land. Sheep got their features after many centuries of life on certain soils; but in the stocking of the country, it is only in recent years that this has been recognised sufficiently to bring about changes, and it is still not recognised as much as it should be.

A list of the more important breeds is given here:

Breed.	Wool—long, medium or short	Colour of face.	Habitat, or conditions under which they are most often kept.	Horns.
Southdown	Short	Mouse brown, not slaty or sooty	Downs, grass, arable	None
Hampshire	Short	Black to mouse colour	Downs, but chiefly arable	None
Shropshire Down	Short	Black	Good grass and arable	None
Oxford Down	Medium	Blackish, not speckled	Chiefly arable	None
Suffolk Down	Short	Black	Chiefly arable and heath	None
Clun Forest	Short to medium	Mottle tan and white	Hill and slightly arable	Occasionally

Breed.	Wool—long, medium or short	Colour of face.	Habitat, or conditions under which they are most often kept.	Horns.
Kerry Hill	Short to medium	White and black speckled	Hill and slightly arable	Occasionally
Cheviot	Short	White	Hill and grass	None
Welsh Mountain	Short	White, tan, varying	High hill	Strong
Softwoolled Welsh	Short	White, varying, pink nose	Hill	Occasionally
Blackfaced Scotch	Long, hair-like	Black & mottled	Grass and hill	Strong
Dorset Horned	Medium	White, light nose	Grass and some arable	Strong
Dorset Down	Short	Grey, type not quite settled	Down, arable and pasture	None
Wicklow (Irish)	Short	White	Hill	Slight
Kerry (Irish)	Short	White, pure, nearly extinct	High hill	Moderate
Devon	Long	White	Grass	None
South Devon	Medium to long	White	Grass	None
Dartmoor	Medium to long	Whitish	Moor and grass	Slight
Exmoor	Medium	Whitish	Moor and grass	Curly
Herdwick	Medium	White, with brown specks	Hill and grass	Rams generally
Lonks	Long, hair-like	Black and white	Hill and grass	Short
Gritstone	Long to medium	Black and white	Grass	None
Cotswold	Long	White to grey	Grass	None
Leicester	Long	White, rare spots	Grass	None
Border-Leicester	Long	White	Grass	None
Lincoln	Long	White	Grass (and to some extent arable)	None
Roscommon(Irish)	Long	White	Grass	None
Kentish or Romney Marsh	Long	White	Grass	None
Wensleydale	Long	Blue or slate	Grass	None
Ryeland	Medium	White	Grass	None

These features are given as a general guide, and they do not make narrow distinctions.

In respect to more detailed features the reader is referred to the several breed societies, or to the National Sheep Breeders' Association.

The longwool sheep, the white-faced sheep of the rich plains, are quite distinct. In the greater number of cases the short wools are heath breeds, though a few of the sheep of the higher hills are of older origin. The fact that sheep are existing on high hills does not prove that they are distinct from the heath breeds found at lower levels; they have merely acquired features and habits which grazing on rougher land at higher elevations develop. Heath breeds, the foundation of so many of the modern shortwool breeds, were generally, but not always, horned; some had white faces, and some dark or mottled, some tan—in fact, several intermediate shades between white and black, and illustrations of these varied colourings were found indiscriminately scattered about the country. They were found on the high hills, heaths, forests, broken woodlands, moorlands, and even on some of the better class lowlands, particularly where rough land was contiguous or nearly so. There was a common relationship between them, but environment exercised striking influence over them. This influence, mainly of soil and its herbage, but also climate, is very frequently shown to-day by the variation in type that takes place when animals of the same parents even are sent to different districts ; whilst the change in a few generations is most noticeable. Breeders who establish flocks away from the indigenous home of a breed are always placed at a disadvantage in maintaining the accepted breed features, because the original breed features resulted from the continuation of conditions which had been brought to bear on them through centuries. A breeder in a district indigenous to a breed does not require to bring in fresh blood to retain the breed features, but the man at a distance frequently has to do so or lose type. Sheep become lighter or darker in the face, the wool piles closer or looser, the constitution changes ; in wet climates the wool becomes more open and longer ; in dry ones closer and shorter ; the sheep increase or decrease in size— in fact, undergo general change in accordance with the degree and nature of the changes the sheep have to meet.

At first sight the Southdown, the old horned Norfolk (of which some flocks are still extant, the foundation stock of the Suffolk), and the blackfaced Scotch mountain breeds, seem very dissimilar ; but in their unimproved state they were far more closely alike ; one has to clear one's mind of the appearance of the sheep to-day, after they have been remoulded by the breed maker, or one is liable to make great mistakes. But shear an old Norfolk ewe and a Scotch mountain ewe as they now are, and very little difference can be observed. The old Norfolk, after centuries on the poor heaths of East Anglia, was ill-shapen, long-legged, a slow feeder, and very short in the fleece. In that dry country a short coat protected the sheep enough. On the bleaker, wetter Southdowns the fleece was longer and thicker, but owing to the greater richness of the feed, particularly because of the many legumes found in it, and which were almost wholly absent in the Norfolk heaths, the sheep had to hunt less for their food, and were shorter on the legs, and in mutton yield far superior. There was, however, another reason. A very common error prevails in the popular mind that the Southdown is entirely the product of the chalk heaths or downs of Sussex. Before root culture came into vogue, and even after, the sheep of the Southdowns were regularly wintered on the Weald ; the cold, wet clays of this grass district carried an altogether different but useful herbage.

In support of this statement I have held the sheep memorandum and account book of Mr. Boys, of Eastbourne, in which were recorded all his sheep and wool transactions from 1742 to 1783.

His sheep went out on to the Weald regularly every autumn, and spent half the year there, and the names of those with whom he agisted them are well known among the Sussex farmers of the present time.   This accounted very largely for the Southdown being the most developed of the heath breeds when Ellman started to improve them, and also for the reason that the Southdown mated so well with all other heath breeds.   The Southdown was far ahead of the other breeds of the chalk heaths or downs at that time.   The Scotch Blackfaced Mountain Breed used to be called the Blackfaced Heath Breed.   As a heath breed it doubtless gradually found its way to the hills of Peak and Pennine range, and worked its way north into Scotland, acquiring more mountainous habits and developing a fleece better suited to resist the heavy rains and exposed feeding grounds.   There is evidence of this in older writings, though there is a tradition that they are derived from a shipwreck during the time of the Spanish Armada.   Those who have followed the history of our live stock must have been amused at the number of breeds to which this tradition attaches. Had it been a Noah's Ark instead of a fighting fleet it could scarcely have distributed breeds more successfully, whether it be of horses, cattle, or sheep.   Moreover, there is rarely anything to suggest Spanish origin in any of the animals ; but failing a knowledge of the origin, the Armada has been a convenient source to suggest. Yet whilst the Norfolk breed existed as a horned breed, the Southdown had gradually lost them, though a coarser, but closely allied, heath or forest breed, of which there remained some until comparatively a few years ago, living on land adjoining that which the Southdown frequented, and known as the Ashdown Forest breed, probably the last of the breeds of the great Wealden Forest, which had no chalk land, carried horns or snigs.   The Southdown was in all probability only a section of the Wealden breed which found its way to the chalk downs, and gradually developed better features.   For those who are not acquainted with Sussex, it may be stated that the term down is not confined in its use to the chalk hills, but is conferred on any smooth, rounded rising ground. Pilt Down, from which Mr. Dawson took the ancient human skull recently, is a considerable distance from the chalk hills.

These three breeds, having a common origin a long way back, show how environment influences type, and how necessary it is that it should—for the fleece of the old Norfolk would be of little defence in wet, exposed hills—and it is not reasonable to suppose that the Blackfaced sheep evolved its fleece to meet conditions all at once.   It gradually developed some other excellent features which are proving highly useful to-day ; but its remote origin from heaths or forests (a few centuries ago a very large portion of the country was in this condition), together with the great constitutional vigour its harder life has evolved, make it suitable to come from the highlands on to the lowlands, as it is doing with such great success in many parts of England. Had it been of the older type of mountain sheep there is little evidence to suggest it would have been nearly so successful as it has proved. For although there are a few instances where breeds ascribed to the old mountain races have become useful on some low lands, this has not been brought about except where much other blood has been brought in, whereas the Blackfaced sheep is singularly free from alien blood.

However, it is seen that blackfaced heath breeds of common origin change greatly under different conditions. The gradations in colouring to white are simple, the blackfaced Scotch is not an all-black, and in our pure breeds—whether of cattle, sheep, or pigs—colour is very much a matter of selection during the past century and a-half; on some soils colour disappears until a white-faced breed is found; but the great family of heath or forest breeds came from an original stock. This accounts for much of the success which comes in crossing to produce new breeds, or for crossing with pure breeds to make desirable crosses to stock land which may have had the indigenous breed pushed out by imported blood when the first great move in sheep improvement prevailed, but is not really well suited with a breed, although the breed holding the land may have been accepted as the correct one to keep for fifty or more years. This gives an explanation why the dictum that a man should stand by the breed he finds on the land does not always hold good. Generally where a modern breed is kept where the original stock was indigenous to it, reasonably good, and sometimes excellent, results are obtained; but where the modern type is an imported one—that is, has no relationship to the truly original stock—it is highly probable that the modern stock does not give such satisfactory results. It is in these districts that trials should be made with other breeds to find something more suitable. It is why so much change in breeds is taking place in certain districts. But all changes need to be made with due regard to the systems of farming, and the food available, as well as to the suitability of the mutton for the available markets, whilst many other minor considerations must be regarded.

What has been said suggests the value of indigenous breed characteristics, though when sheep go to other conditions the sheep will at once try to adapt itself to its new surroundings; where the new conditions are widely different to those to which the sheep have been accustomed, they will break down and become unprofitable before they can adapt themselves. For the good of breeds it is of first importance that the flocks kept on the indigenous ground should retain the indigenous features; if not, there is the risk that, in course of time, the true stock would be lost.   Breed societies should very jealously guard this point, even though there may be some features in which from appearances a sheep reared on other land may be superior ; the danger is never one that is likely to be brought about very rapidly, yet it may creep in.

But there is another aspect in which to regard a breed.   It is the boast of some that no other race was ever put on the sheep which were the indigenous foundation of the breed ; that is, since Bakewell set going sheep improvement, no other blood reached the original stock.   This is very hard to prove, and in view of the great amount of crossing that took place before sheep were handled on definite breed lines, very hard to believe.   Moreover, for the general usefulness of a breed, that is, for its suitability to go to improve stock under a greater variety of conditions, increase in the number of breeds which went to compose it—provided they are properly fixed—give it a wider field of usefulness.   Narrow breeding, like narrow views, can only fit in in narrow places.   A composite breed possesses the indigenous characteristics of more than one type ; some breeders could not receive a greater insult through their flock than by being told that there had been an importation of blood at some time ; they do not seem to recognise that this would imply a narrow set of features that adapt it to a small area ; whereas, the more breeds well incorporated and fixed, that go into the composition of the breed, the better are the chances that it will mate well with others, and accommodate itself to changed surroundings.   But the breed must have fixity of type, or mongrelism will arise ; therefore a pedigree, that is, a record of breeding, to show that the race has existed without the admixture of strange blood over a considerable number of generations, has value in giving this assurance.   When a pedigree records a long period of careful breeding and selection on skilled lines, it is of great value ; but mere pedigree may have little value if the principles of breeding have been ill carried out ; it is then merely a record of the perpetuation of features that ought to have been eliminated.

There is no doubt that, apart from good management, the soil of a particular district or farm will produce better sheep than will other farms even near by, no matter how skilled the breeder, and it is on these specially favoured farms that the breed type is established ; if there has been good and enlightened breeding on one of these farms, the characteristics of the breed are established from it, and pedigrees from such farms certainly have value.   But breeders must produce the class of sheep for which there is a call. Very often the call is for a sheep better suited for countries over the seas, and then the home breeder has to decide what will best suit him.   Several breeds are undoubtedly suffering in their home sales because the breeders have followed features which the foreign buyer has specially called for. Another point which requires more recognition is that as a breed is taken further from the conditions indigenous to it, as where food is produced on the land of a different character to that which developed the indigenous features, the sheep will tend to alter, being influenced by the new conditions. Sheep for centuries fed on grass are sure to modify when roots and other cropping form a great portion of their food, and when, for many generations, they have in addition a large quantity of cake or other concentrated food, they are likely to lose more of their indigenous features. As all breeds have intermixture of blood, the sheep will tend to follow the features of the component breeds which appreciate or are most influenced by the change of food. Breeds ought not to stand still, but they need to be steered carefully. The breeder has not merely to think how it looks, but rather to consider what it is and what it will do.

For some years special attention has been paid to the principles of breeding set out by Mendel, and Professor Wood, of Cambridge, carried out some very interesting experiments on sheep, showing how it was possible to cross breeds of great antagonism in feature to breed new types giving them fixity of feature, and to a considerable extent of characteristics. However, breeding must be a work of a very exhaustive nature if an attempt is made to take full advantage of new principles—and time and opportunities have not offered this at present. With such new breeds it has yet to be discovered how the indigenous forces of soil, climate, elevation, food, &c., will modify them in the course of years. The destruction of cells controlling certain growths may, as far as can be reasonably argued, be so completely effected that they cannot be redeveloped whilst breeding in true lines is followed, and thus objectionable features may be eliminated; but at present we lack knowledge as to how the unaffected cells would be influenced. They might or might not be affected unduly by natural surroundings, though there seems no reason why selection should not keep these corrected, as such things are controlled by selection as it has been carried out hitherto. One can see considerable possibilities in its connection with sheep breeding and improvement, but apparently what is done until the principle is subjected to much more investigation will have to be regarded as experimental.

CHAPTER IV.

SHEEP CENSUS.

In 1908 the Boards of Trade and Agriculture co-operated to ascertain the number of sheep in the several breeds of sheep in Great Britain. The figures arrived at may be accepted as being fairly representative of the breeds and types; but the Boards recognised the disadvantages they were working under in taking on the work with little opportunity to make arrangements for obtaining absolute exactness. For practical purposes they may, however, be accepted as a reasonably reliable guide. The divisions of " Scotch," "North " or " North Country," and " Downs," are not actual breeds, and would be returned by those who are hazy as to the actual breeding, and therefore describe them in a generic manner; probably a large portion would be half-breds. Possibly the divisions would have been better expressed as " breed and crosses in which the breed predominates." The large number of Blackfaced Mountain Sheep would doubtless surprise many, but probably the big head of Cheviots would be a greater surprise; one would expect the popular verdict to be that Welsh Mountain Sheep would be found in greater numbers.

The influence of the Southdown, as shown by the pure Southdown and the breeds which have acquired characteristic " Down " features, falls little short of the Blackfaced Mountain in numbers—5,278,000 against 5,579,000. Further, the greater size, and more rapid maturity of the Downs, are points of great economic importance; and as they so quickly reach the butcher, the sheep under Down influence undoubtedly contribute the greatest weight of mutton, although probably a large proportion of those included under the heading of Scotch may justly be classified as Blackfaced. No doubt the popular estimate of individual breeds has been largely influenced by the extent to which they have appeared in the showyard, and the publication of this analysis of breeds serves useful purposes. One or two other points may be referred to. The sheep commonly classified as Longwools are represented by 4,840,000. The Cheviot, which stands second on the list, has had a share in the making-up of some recognised breeds, and these together give a total of 4,000,000, without including breeds which have been partially improved, as a section of the Welsh Mountain. In the aggregate the sheep classed as Mountain races exceed any other type, which is not unnatural, as sheep are essentially the animal of the high-lying pastures and heaths ; as a rule these have been less directly crossed than have other breeds, which follows from the fact that where the food is little altered by man's agency it is not advisable to disturb Nature too much.

The Census, as arrived at, gives :—

Blackfaced Mountain	5,579,182	Devon Longwool	750,688	"Downs"	145,920
Cheviot	2,650,817	Leicester	676,566	Chun Forest	119,285
Welsh Mountain	2,600,131	Radnor	654,547	Lonk	113,613
Lincoln	1,850,074	Herdwick	531,457	Dorset Down	99,853
Hampshire Down	1,672,340	South Devon	353,826	Shetland	79,756
Shropshire (Down)	1,603,879	"North" or "North Country."	302,599	Ryeland	28,936
"Scotch"	1,173,663	Wensleydale	259,450	Cotswold	26,966
Oxford Down	1,082,737	Border Leicester	231,786	Limestone	12,199
Kent or Romney Marsh	1,044,569	Dartmoor	199,475	Other breeds or descriptions	745,105
Suffolk	918,034	Dorset or Somerset Horned	179,598		
Southdown	755,389	Mashams (Cross).	173,005		27,119,730
		Exmoor	172,347		

CHAPTER V.

DOWN AND HORN BREEDS.
SOUTHDOWN.

The Southdown, the first of the heath breeds to be brought under systematic improvement, originated on the Sussex chalk hills, which are high-lying heaths carrying exceptionally sweet herbage, in which the minor legumes and aromatic plants are very prevalent. These hills attain the height of 800 feet at the highest point, but are generally considerably lower. The term down, in Sussex, is applied to many other hills which are not on the chalk. At the base of the chalk downs there is generally an outcrop of Greensand forming rich land and excellent lair for sheep; in addition to this there is often light alluvial soil before the strong land of the Weald is met; rich marshland is near by in some parts, especially towards the east. These are the soils on which the Southdown sheep was bred and developed over many centuries; but one essential feature in the development of the Southdown breed not commonly regarded was the fact that it was customary for centuries for the Down breeders to send their sheep to agist on the strong land of the Weald from October to April. I had possession of the sheep account book of Mr. Boyes, of Eastbourne, in which there was a record of every sheep in his flock of 800 from 1742 to 1783. (I gave this memorandum to the National Sheep Breeders' Association.) The sheep from this flock were of good quality for the time, and Mr. Boyes was a personal friend of Mr. Ellman, of Glynde, who brought the breed into prominence by his splendid skill as a breed maker. From this account it can be seen that the flock was sent yearly to farms in various parts of the Weald to winter, evidently as part of the ordinary practice of the Down breeders before turnips, mangels, and other crops were grown in sufficient quantity to provide winter food at home. It is interesting to note the price of sheep at that time. In the first year of the record, the main sale of lambs at Lindfield Fair on July 25 was made at one shilling and ninepence halfpenny per lamb! And sheep a year older were sold for eight shillings, which was considered satisfactory. Next year sheep were dearer, and the lambs fetched two shillings each. Gradually the price increased as the influence of better farming and the improvement made on the breed by Ellman was felt; and in the last year of the record the lambs at Lindfield Fair made over eighteen shillings—which may be regarded as a truly wonderful improvement.

Mention has been made of these points dealing with the origin of this historic breed, which has had such great influence in the making of so many breeds. In recent years, and for a long while, the breed, so far as Sussex is concerned, has been mainly kept on the chalk downs, but it got its many indigenous features through the sheep spending their lives on such a variety of soils, and doubtless it was because its characteristics and constitution were so built up that it crossed so successfully with other heath breeds, which subsequently partook of these features, and are known as Down races—of which the Shropshire, Hampshire, Oxford, and Suffolk are the more notable, though there are other breeds which directly or indirectly have been influenced by the Southdown. It has been the great improver of the heath breeds. Another common error is that because it comes from the south of the country it is necessarily a tender breed ; as a matter of fact it is wintered on very cold and exposed hills, and has to take a considerable amount of exercise in grazing, although it may be folded in by no means sheltered pens at night. The Southdown is not a tender breed, but to maintain its typical breed characteristics it requires to be kept on the chalks, or in course of time they break away, although the breed will grow well and fatten readily on other land when placed on it for fattening purposes. The other races with which it had so much to do in the making would not have been as vigorous as they are known to be if the Southdown had naturally a weak constitution ; but it is essentially a sheep of good land and sweet food, just as the Shorthorn is among cattle. On chalk land at a distance from the Southdowns it thrives well, but is difficult to keep to the recognised breed type—but it is an excellent animal all the same. It has long been associated with arable farming and close folding on roots, and fattens well on them ; however, the ewes in Sussex are not put to the ram early as a rule, consequently many of the lambs are run over the following winter to be fattened out late in spring or during summer. A considerable number is fattened as fat lamb, and if well done throughout, they may be fattened at fair weights by Christmas.

The work of John Ellman, of Glynde, in bringing this rough heath breed into such good form ranks among the finest achievements in British farming, and was the foundation of the high quality meat obtainable throughout the world. In the making of any breed where the foundation stock is a heath type, it is practically an essential that some dash of Southdown blood be infused ; it may be in greater or less degree according to what is required, but if only a very small infusion be made, the thriving powers are improved with no loss of quality to the meat But, as in most breeds, those who have greatest influence in breeding need to be careful not to encourage the development of too much fat in the pure-bred ; however, there is one feature about Southdown mutton, few are so fastidious that they refuse to eat all the sweet and dainty fat of this great little breed, unless the animal be foolishly overfed. If Southdown mutton costs more per pound than other kinds, those who habitually eat it regard it as not being more expensive, as there is absolutely no waste. Owing to the hard life these sheep lead on the exposed downs, a high average number of lambs is not raised ; but on richer land, where they are done well, they are equal to producing a lamb and a-half-average. The Southdown gives a close fleece of fine wool ; when Ellman took the breed in hand 2 lb. to 2½ lb. was about an average. A very small infusion of Leicester blood was put on the sheep in the early days, and the practice lasted a very short time ; but probably that little gave timely help to the breed improver in improving the fleece. It is scarcely likely from analogy with other heath breeds that selection would so rapidly have brought about the change, unless help had been given from outside sources. But all traceable influence has long since gone, as is natural when there has been no introduction for more than a century. Apart from its influence in breed making, the Southdown has great value for crossing with other breeds, and it is a favourite cross with the Kentish or Romney Marsh breed, which gives a sheep of fair size, producing mutton of good quality. The cross makes good, fat lambs, or they may be run on as wethers ; these crossbreds do well in the turnip fold, but they can be very cheaply kept as a purely grass sheep, and well suit the country lying to the north of the Southdowns and the Kentish marshes.

The Southdown is primarily a mutton breed, but it cuts a close fleece of wool of very fine texture and quality, ranking highest in the British wools, running from 5 lb. to 8 lb., according to breeding and management of the flock. It is to the Southdown that the breeds influenced by it thrive so well in the close fold ; it was Ellman's intelligent farming that developed a wide system of catch cropping and close folding that enabled him to feed and maintain an improved breed ; this aptitude for the close fold was transmitted, with the result that all Down breeds are well suited to arable land farming.

The description and scale of points recommended by the Southdown Sheep Society are given below. Somewhat similar descriptions are given where other breed societies have published them. They are useful as defining the views of those who have the direction of the several races and the relative value they attach to the individual points, whilst the features held as desirable for disqualification hold special value in the breed's interest. Judging strictly in accordance with scale is rarely practised ; in fact, whatever the animal, strict point judging cannot be followed wholly satisfactorily. This is met with in the case of the Southdown to some extent by the points allowed for under " general character and appearance," and probably judges often subconsciously in their awards give more than the actual points set down. An animal may be good on points at either end, but they may suggest two different types of sheep, in which case the scale of points can be misleading ; therefore, there needs be a wide margin left, and in practice it avails.

Description and Scale of Points.

General character and appearance..   10

Head.—Wide, level between the ears, with no sign of slug or
       dark poll   8

Face.—Full, not too long from the eyes to the nose, and of one
       even mouse colour, not approaching black or speckled,
       under jaw light   4

Eyes.—Large, bright and prominent   2

Ears.—Of medium size, and covered with short wool   ..   2

Neck.—Wide at the base, strong, and well set on to the
       shoulders, throat clean   5

Shoulders.—Well set, the top level with the back   ..   7

Chest.—Wide and deep   5

Back.—Level, with a wide flat loin   10

Ribs.—Well sprung, and well rubbed up, thick through the
       heart, with fore and hind flanks fully developed   ..   7

Rump.—Wide, long, and well turned   4

Tail.—Large, and set on almost level with the chine   ..   4

Legs of Mutton.—(Including thighs) which should be full, well let
       down, with a deep wide twist   10

Wool.—Of fine texture, great density, and of sufficient length
       of staple, covering the whole body down to the hocks
       and knees, and right up to the cheeks, with a full fore
       top, but not round the eyes or across the bridge of the
       nose   10

Skin.—Of a light delicate pink   5

Carriage.—Corky, short legs, straight, and of one even colour,
       and set on outside the body   7

                                                    ———

                                                    100

Disqualifications.

The following are reasons why judges should not, at breeding stock shows, award a prize to otherwise good sheep :—

(a) Horns, or evidence of their presence.
(b) Dark poll.
(c) Blue skin.
(d) Speckled face, ears and legs.
(e) Bad wool.

SHROPSHIRE.

The Shropshire was the first recognised evolution resulting from the crossing of a native heath breed with the Southdown; in other words, it first received Royal status as a Down breed, although the county with which it is associated is far distant from the chalk downs; it, however, illustrates very clearly the prepotency of the Southdown on allied breeds, and the aptitude of the resultant breeds to maintain Down characteristics when away from the indigenous factors of the Sussex Southdowns. The breed more directly sprang from animals which were indigenous to some of the richer tracts of land in Shropshire and Staffordshire, of which the Morfe Common and Cannock Chase are respectively typical. The native breeds of both sheep were small, horned, and relatively unthrifty before the Southdown was brought in. More or less indiscriminate breeding went on for some time, as it did with all breeds when they were in the melting-pot a century or so ago, when the two great early-improved races, the Southdown and the Leicester, were influencing the sheep throughout the country. Shrewd men on the best sheep land noticed that their sheep were better than their neighbours' on less suitable land; and in that early wave of enthusiasm paid attention to breeding as their forebears had not. Gradually the crosses, by selection, began to take type, and, where skilfully handled, laid the foundation of new breeds. The Shropshire emerged in this way.

Mr. Meire and Mr. Adney were specially notable as early breeders, though they adopted rather different lines of breeding, and Mr. Meire bought or hired freely Southdowns from Mr. Ellman; and Prof. Coleman, writing on the sheep nearly forty years ago, stated that Mr. Meire introduced the Leicester blood with great judgment, and then fixed the type by close breeding. In 1853, at Gloucester, Mr. Meire and Mr. Foster secured all the prizes from sheep descended from Mr. Meire's stock. Mr. Smith, afterwards highly successful, founded his flock on Mr. Meire's, and later both he and Mr. Foster used the Oxfordshire Down. Lord Chesham, who was so successful at the end of the 'seventies, purchased a good many sheep from Mr. Smith, so a considerable percentage of Meire's blood must have existed in his sheep. The 'seventies brings us to the time when Mr. Mansell and Mr. Evans made their mark on the breed, and did much to develop modern type. How well these selectors handled the several types involved in bringing the thriftless native sheep to the excellence it has attained in modern times needs no comment, but it stamps them as breed makers of first excellence. By the use of the several bloods they incorporated they have produced a breed which has not only greatly influenced the sheep industry of this country, but has made its mark on the flocks in Scotland, and very much in Ireland ; whilst in the big sheep countries over the water, where wool and meat have been considered, the records of the Shropshire sales is sufficient evidence. The Shropshire stands as an excellent example of the value of combining the blood of several breeds to produce a cosmopolitan build, and there is no testimony so valuable to a breed as that of its cosmopolitan influence.

The breed is met with in a pure condition in Staffordshire, where the Cannock Chase sheep was used as foundation stock by Mr. Keeling, who established his flock in the early 'thirties, though Mr. Coxon, of Whittington Heath, who started his flock in 1825, did much for his county's sheep during the forty years he handled it. But it is almost safe to say that nearly the sheep world over the Shropshire can be met with. Pure wool and meat has been the watchword of the Shropshire men, and if one's personal idea is that for home purposes wool has been a little too much sought after at the expense of the lean meat, there are probably those who would refute this ; but the success of the breed abroad has warranted the breeders' methods. In the personal view that excessive covering of the face with wool tends to promote the fleece at the expense of lean meat one feels satisfaction that those responsible for the breed have sent out an edict, that this covering shall not be held in the regard that it has possessed for some years.

If very much a pasture breed, and in first instance the sheep of the redlands, as the Southdown is of the whitelands, the Shropshire does well in the fold. It has had considerable influence on the forest and hill breeds on the land adjoining, and crossed with these very early and excellent lambs are produced. Speaking broadly, the Shropshire is a sheep which does better north of the Thames than south of it, though in the cross I have seen it do well not far from the English Channel. The breed is remarkable for its compactness, and for meeting the points which generally appeal to the eye of the sheep lover.

The Shropshire in the 1908 Census was returned practically in the same numbers as the Hampshire ; had the Irish figures been added, it is highly probable that the Shropshire, with its crosses, would have stood relatively higher.

The Shropshire Sheep Breeders' Association and Flock Book Society was formed in 1882, but it does not insist upon a scale of points for animals in competition. Mr. Alfred Mansell, whose interest in the breed in every direction, has been of incalculable value to it over many years, and whose influence has been highly beneficial on all breeds, wrote me to the effect that the Association preferred recommending the distinguishing points, thus giving breeders a wider discretion and preventing the narrowing effect of point judging. The more important are as follows: " The head and the face (which should not be too long) should be completely covered with fine white wool; the face and legs a nice soft black in colour; the ears also dark and small or medium, fleece very dense, fine, and of medium length; the body and legs covered with an even quality of wool—coarse wool about the thighs, or light and thin wool on the shoulder points being a great fault, as also are patches of black or grey wool. The skin is pink, and free from blue or dark spots, the body square, on short, straight, stout legs, with good bone."

The Hampshire Down.

The Hampshire Down is the sheep of the Western chalk heaths or downs. It is a true Down, and has spread from its indigenous chalks to those further afield, and keeps its character well in them. It is one of the composite breeds, in which the Down characteristics were derived from the Southdown. In its composition the native Wiltshire white-faced horned heath breed may be regarded as the indigenous stock, and the bold Roman nose remains as absolute evidence of its presence. The Berkshire heath breed, the Berkshire Nott, or Knot, or hornless sheep with a black or only slightly spotted face, associated with that county, considerably entered into the race, and is probably the source of the black face and leg colouring, though it is known that the Southdown, on going to some districts, has a tendency to get darker in the face. As its geological position indicates, doubtless the more eastern position of the breed was more influenced by the Nott than was the western; and in support of this the breed, as found in Dorset, are much lighter in the face. The Cotswold also was brought in as an early cross, and through him a touch of the Leicester. With all this intermixture the Down characteristics have been evolved, and, in the opinion of many, the handsomest of all the Downs.

The Hampshire Down has the features of early maturity well developed, and it is doubtful if any breed of Down characteristics will, in open field treatment, come to the butcher at heavy weights so quickly without showing an undue tendency to fat. It is, however, a breed that tends to the appreciation of good living, and over large numbers no breed is so well provided with suitable food to bring it along. The wide system of catch cropping which Eliman evolved to suit his requirements when improving his Southdowns has been adopted, and to some extent elaborated, by the farmers on the Western Downs, with the result that they are able to wean their lambs very early in the year; in fact, January is the great lambing month with a large proportion of the flocks.

This early lambing, and the good and varied feeding that the lambs receive until the August sales, when many are sold to be wintered on roots in distant counties, ensures such growth that, combined with the features of early maturity that so mark the breed, it is possible to turn out heavy sheep by November and through early winter. On the root land in the Home Counties and South-east Midlands, as well as further afield, the Hampshire is exceedingly popular in the fold. It is also largely used for crossing with longwool breeds, especially where they are required to be sold in the northern markets. If anything, the Hampshire has developed a tendency to make rather deep fat when fattened right out, and with its big weight it has not quite retained the position it held in some districts a time back. Much good doing, and selection to cover the face with wool which has predominated over other good features in the past quarter of a century, are probably accountable for some of this; though, for fat lamb purposes, it is not so objectionable. Many good breeders view this with some anxiety, and are casting about to remedy the position, especially as some smaller breeds are encroaching on ground that was held by the Hampshire. The Hampshire grows to heavy weight if required, and cuts a good fleece well free from black hairs; with longwool breeds it makes a good wool cross. How far sacrificing the tendency to produce lean meat should be indulged in favour of the fleece should be considered by those interested in any breed.

There is no doubting the fine constitution and vigour of the breed; for, of all breeds, the Hampshire not only gives the impression of these, but is the model of a thick-fleshed sheep, which is the more notable when the thin flesh of the original stock is considered. In spite of the criticism on one or two points, the Hampshire is a grandly-fleshed sheep, and, under ordinary farming conditions, fills an important place in the supply of high-class mutton in early winter, when other Down breeds are less available. Pure, or with a slight dash of Oxford, it is the mainstay of the folders on roots in Herts, Beds, and contiguous counties, and its aptitude to fatten whilst it grows is well marked. Relatively few Hampshires are kept as wethers, as their early maturing powers enable their being sent to the butcher whilst under twelve months old—very few survive to meet a second shearing. Well treated, on ordinary arm practice, the tegs will go out at 8 st. to 9 st. in November, but under the high pressure this breed can stand greater weights can be obtained. The ewes clip about $4\frac{1}{2}$ lb. to 5 lb. of wool, and shearlings 6 lb. to 8 lb. As the treatment of this breed is dealt with somewhat fully in another section, under the heading "Management of a Breeding Flock on the Chalks," further details need not be given here.

" The Standard of Excellence," and the " Objections," as outlined by the Hampshire Down Sheep Breeders' Association, are as
follows :—

STANDARD OF EXCELLENCE.

Head.—Face and ears of a rich dark brown—approaching to black—
well covered with wool over the poll and forehead. Intelligent bright full eye. Ears well set on, fairly long,
and slightly curved. In rams, a bold masculine head is
an essential feature.

Neck and Shoulders.—Neck of strong muscular growth, not too
long, and well placed on gradually sloping and closely
fitting shoulders.

Carcase.—Deep and symmetrical, with the ribs well sprung, broad
straight back, flat loins, full dock, wide rump, deep and
heavily developed legs of mutton and breast.

Legs and Feet.—St ongly jointed and powerful legs of the same
colour as face, set well apart, the hocks and knees not
bending towards each other. Feet sound and short in
the hoof.

Wool.—White, of moderate length, close and fine texture, extending
over the forehead and belly, the scrotum of rams being
well covered.

Skin.—Pink and flexible.

OBJECTIONS.

Objections.—Snigs.
White specks on face, ears and legs.
Thick coarse ears.
Black wool.
Coarse wool on breeches.
Protruding under jaw.
Excessive loose skin under neck.

THE DORSET DOWN.

The Dorset Down has been instituted as a distinct breed in
quite recent years. It is in reality an offshoot of the Hampshire, and still holds many points in common. Its future type
cannot be said to be definitely defined at present. In Dorset
sheep have a tendency to develop white or light faces ; but the
Dorset men have all along fancied a smaller and lighter faced
sheep than has been popular with the men to the east of them, and
have favoured something more after the Southdown type. Their

sheep were never so much impressed by the Berkshire Knot, which is an additional reason for their being of lighter type than the Hampshire. The late Mr. Saunders, of Watercombe, who was acknowledged to have done much to improve the Hampshire Down, and, in his day, was a successful exhibitor, held out against the dark face, and in following his lead there is reason to believe that Dorset men are acting soundly in aiming at a sheep which naturally follows the features which its soil and climate promote. The breed is being modified, but an outsider may think that to be distinctive rather more boldness is desirable, and that something more than a rather smaller breed with a rather lighter face should be aimed at, but it is obviously unfair to criticise at this stage. As it is, the chief distinction between it and its allied breed is that it shows more of the Southdown character.

That it is a good type of sheep, and that the mutton is very saleable, cannot be denied ; moreover, the wool is very clean, all of which are strong factors in a breed. It is satisfactory that those who are showing the breed are very enthusiastic and energetic in their work. Whether the Cheviot, Exmoor, Ryeland, or other high-class small breed could not be imported with advantage, and a more thorough re-casting be adopted, is a suggestion that must come to the mind of some of those who are cognisant of the breeding results being obtained in other places. The first aim in a breed is to make it suitable for its district, but breeders of high-class stock look for a more extended market, because, generally, it is only in this way that great profits are obtainable to allow them to go to great expense in experimental breeding. That the breed, until it is more drastically treated, will carve out for itself a following far from the district of its origin is doubtful. That it is an excellent foundation stock to originate a breed in association with some of the good minor races with special mutton properties, and not too much fat, there can be little doubt.

THE OXFORD DOWN.

The Oxford Down, in its early days, was known as the Down-Cotswold. Broadly speaking, it resulted from crossing the Cotswold with the Hampshire, though Mr. Milton Druce, writing on the subject of its origin some fifty years ago, thought that there might have been a little help from the Southdown direct, though he leaned to the view of the simple cross. From the cross a very useful sheep has been evolved ; the largest with distinct Down characteristics ; popular on some soils as a pure breed, and in good demand in others for crossing, carries a heavy fleece, and begets lambs well suited for the early fat lamb trade. It crosses back well with the Hampshire, and on cold winter lair or arable land no sheep beats this cross. It may be regarded as the typical breed of the Oolites, where other Down breeds do not do their best ; for without some pronounced proportion of Cotswold blood these soils are too chilly in wet folds for sheep not well protected against mud.

This feature of resisting cold doubtless accounts for the large number of Oxford Down rams which yearly go to Scotland. It is largely used on longwools when it is desired to maintain or increase size and give Down quality to the mutton. It has rather too great a tendency to produce fat on the back to compete in price with some of the finer quality Downs, but it comes to great weight quickly, and on some soils is a decidedly popular sheep to keep. It is of a more distinctive Down character than it was a quarter of a century ago ; the Cotswold forelock has to some extent been modified in the endeavour to imitate the head covering of some other Down breeds ; in fact, one felt well certain a few decades ago that some flockowners had hastened the advancement of the breed by further recourse to other Down breeds. On the whole, this has had a good result, and at no time in their career have these sheep been so strictly typical. The breed is very popular in crossing with certain longwools, notably with the Lincoln, in districts where there is still a ready sale for big mutton. It is an impressive breed, and its influence on longwool mutton is very marked, greatly adding to its value and popularity by producing the Down flavour and removing the more tallowy element in the fat.

The breed is in considerable demand for crossing with several breeds to get fat lamb, as good quality and big weights are obtainable. Even when pure, lambs are easily got to 40 lb. to 50 lb. in three to four months. The wool has been much improved in late years, as the breed has assumed more Down-like type ; it is longer than the ordinary Downs if not quite so fine. On the whole the Oxford, in assuming more of the Down type, and a greater trueness to type than it had thirty years ago, has come more into line with modern requirements where a big sheep is demanded.

The Oxford Down Sheep Breeders' Association publish the following description of the breed :—

The *Oxford Down* has a bold masculine head well set on a strong neck, with poll well covered with wool. The *face* a uniform dark colour, the *ears* of good length, the *shoulder* broad, with a *broad breast*, well forward. A full level *back*, the *ribs* well sprung, the *barrel* deep, thick and long, with straight underline. The *legs* are short and dark coloured, standing square and well apart. The *mutton* is firm, lean, and of excellent quality. The whole body is covered with *wool* of close texture, good length, and fine quality.

The Suffolk.

The Suffolk breed, which resulted from crossing the old Norfolk Horned and the Southdown, has made wonderful strides in the past quarter of a century. It was not until 1886 that the Royal Agricultural Society recognised its value as a distinct breed and provided separate classes for it ; the type of sheep has considerably altered in the hands of the undoubtedly skilled breed makers who have steered its course. At the present time it stands first in mutton points among the larger breeds, as indicated by repeated successes in carcase competitions. Its whole appearance is suggestive of " quality," and it provides an excellent example of the fact that head-covering and face-covering with wool are not necessary ; moreover, the clean head, associated with the high class of the meat, the high average increase daily in weight, and the vastly improved wool, show conclusively that head-covering, even in other breeds in which the Southdown has played an important part, when carried to excess is not warranted. It would have been quite easy for the head-covering to be developed on the Suffolk, but the breed makers thought better of it, and have gained all their points without it. They have a breed which does not run to excessive fat, as they avoided that form of selection which is conducive to fat and wool at the expense of the lean meat.

The Suffolk is a most prolific mother, with a heavy flow of milk ; consequently the number of lambs reared and weaned is very big. It is hardy and active, and a good forager ; it is the least liable to foot-rot of any Down breed. The horns, so marked in the old Norfolk, have been bred out. It crosses well with many breeds, and in association with the Cheviot makes excellent mutton, as carcase competitions evince. On the whole, it is better suited as a pure breed to dry lair and a dry climate than to the wetter and colder ; it is raised in the driest climate in the kingdom. Still considerable numbers cross the Scottish Border and some are found in Ireland. If the Suffolk suffers at any point it is in its narrow breeding, the Norfolk and the Southdown being mainly responsible for its formation, and, as such, it is not a typically wet climate sheep nor at its best on heavy land.

The Suffolk, however, is both a grass and arable breed. It is accustomed to graze on poor heaths and also on rich marsh lands ; but it can do extremely well as a wholly arable sheep. Crossed with the Cheviot, an exceptionally valuable sheep results ; and one can scarcely expect that, with this knowledge, breed makers in some parts of the country will not fix a type composed of these two breeds, for it would be very valuable and would adapt itself to a great variety of soils and climates.

The fecundity of the breed is very great, a lamb and a-half

per ewe being no rare average. The Suffolk Society gives shepherds' premiums for the greatest number of lambs raised per ewe. My personal experience with the Hollesley Flock in 1902 was—587 lambs from 303 ewes, inspected on May 1, and I can vouch for its accuracy.

The Suffolk Sheep Society have published the following scale of points :—

SCALE OF POINTS.

Head.—Hornless ; face black and long, and muzzle moderately fine—especially in ewes. (A small quantity of clean white wool on the forehead not objected to.) Ears, a medium length, black and fine texture. Eyes, bright and full 25

Neck.—Moderate length and well set. (In rams stronger, with a good crest).. 5

Shoulder.—Broad and oblique 5

Chest.—Deep and wide 5

Back and Loin.—Long, level, and well covered with meat and muscle ; tail broad and well set up. The ribs long and well sprung, with a full flank 20

Legs and Feet.—Straight and black, with fine and flat bone. Woolled to knees and hocks, clean below. Fore legs set well apart. Hind legs well filled with mutton .. 20

Belly (also Scrotum of Rams).—Well covered with wool.. .. 5

Fleece.—Moderately short ; close fine fibre without tendency to mat or felt together, and well defined, *i.e.*, not shading off into dark wool or hair 10

Skin.—Fine, soft, and pink colour.. 5

Total 100

THE DORSET HORN.

The Dorset Horn breed, having its indigenous home in Dorset and Somerset, is of the heath type, and in its unimproved type had much in common with the unimproved white-faced old Wiltshire horned breed, which was the foundation stock of the Hampshire Down. It is singular in England as having a pink nose, similar to that of the Merino, and old illustrations show great similarity between the unimproved Merino and the little improved Dorset. Whether at any early date the Merino, or other of the pink nosed breeds from the Mediterranean coast line, was brought on to the Dorset is not conclusively shown ; but it is certainly strange that if there were no such intervention that this particular feature should have been developed in this breed. That the breed had something unusual in its composition is strongly indicated by the abnormal periods of breeding, which not only show themselves by the ewes coming into season far earlier than the British breeds, but by the readiness with which they breed again ; in fact, to many, the breed is chiefly recognised through this feature of producing two crops of lambs in a season ; whereas it is only occasionally that a sheep in other breeds will do so.   This habit is un-English, and is probably traceable to the disturbance in season which a sheep coming from a hotter climate might feel.   There is, however, evidence that in the very early days the Leicester and Devonshire Knot were used, but according to Youatt without success, and not sufficiently to disturb the breeding.

The appreciation of the breed, particularly by those in the Home Counties, to which many were sent yearly to breed lamb for early fattening considerably more than a century ago, seems to have stimulated breeders to improve the sheep, with the result that they have long been of good type, although they have shared with other breeds a steady improvement.   It is not merely the great fecundity of the breed, and their suitability to produce lambs for early fattening, either pure or in the cross, but the fact that they attain good weights as shearlings, and cut a heavy fleece of wool of high quality, and provide mutton of good quality that makes them a really valuable race.   The development of this breed by those in some districts not well suited with sheep is worth much consideration.   As a heath breed in origin it has much in its constitution to adapt it to a variety of soils ; it certainly does well on some heavy soils at a distance from its home, as is likely where it has a natural habitat on the Kinmeridge clay, the chalk, and the oölite.   It is a breed quite well suited to supply features in the development or recasting of types to suit modern purposes.

The Dorset Horns are specially liked for crossing with Down breeds to produce fat lamb, for there is no better quality lamb grown when crossed with a good Down.   Many of the sheep are kept on down land, where, in summer, the ewes pasture during the day and return to the arable land for folding at night.   Where they get the advantage of water meadows and other good pastures, they are preferentially placed.   It is customary to put the ewes to the ram about June 20, and the lambing time is generally from the middle of November to Christmas, though some are bred far earlier.   The lambing is generally done on grass, and it is customary to feed the ewes liberally with cake.   The weaning takes place in March, when it is desirable to have swedes, or rye, with mangels available.   The earliest fat lambs are got rid of much sooner.   The lambs cut $2\frac{1}{2}$ lb. to 3 lb. at six months old ; the ewes 5 lb. to 7 lb. ; and the rams 10 lb. to 14 lb.   The wool is very white and clean, and ranks very high in the market.

The Dorset Horn Sheep Breeders' Association has established the following standard of excellence :—

Rams.—Bold masculine appearance, and robust character ; head of great beauty, with strong and long horns growing from the head, well apart on the crown, in a straight line with each other, and coming downwards and forwards in graceful curves as close to the face as may be without necessitating cutting.

Ewes.—Appearance bright, with feminine characteristics. The horns much smaller and more delicate than in the ram.

GENERAL CHARACTERISTICS.

Head.—Broad, full and open at the nostril, well covered with wool from brow to poll.   Face, white with pink nose and lips.

Ears.—Medium size, and thin.

Teeth.—Flat, chisel-shaped.

Neck.—Short and round, well sprung from shoulders, with no depression at the collar, strong and muscular, especially in the ram.

Chest.—Well forward, full and deep.

Fore Flank.—Full, with no depression behind the shoulder.

Shoulders.—Well laid and compact.

Back and Loin.—Round, long, and straight, with well sprung ribs.

Quarters.—Full, broad, and deep, with flesh extending to the hocks.

Ribs.—Well sprung from the back and deep at the sides.

Tail.—Well set up in a line with the back, wide, firm, and fleshy.

Legs.—Well placed at the four corners, straight between the joints, with plenty of bone, well woolled to, and below, the knees and hocks.

Fleece.—Of good staple and quality, compact and firm to the touch.

THE RYELAND.

The Ryeland is a breed which in early times is said to have been of considerable importance in the Western Midlands, though its name is supposed to have been taken from its particular association with the rye lands of Herefordshire. It, however, was an exceptionally small sheep, and was much crossed by the Leicester, when the wave of sheep improvement took place after Bakewell's time ; this resulted in a bigger sheep and some sacrifice of the beautiful fineness of the wool ; for a time the prospects of the breed were poor ; fortunately skilled men took it in hand, and although considerably changed, it has developed into one of the most attractive British breeds.

In view of its type and excellent development it may fittingly be described as the Southdown of the West, although it has no Southdown blood in it, so far as is known. Its popularity is spreading, and although the numbers are small, there is little doubt as to their increase in the near future. It is one of the breeds that can be very usefully employed in crossing or in breed making. Although hornless, the old breed in all probability had heath characteristics, as it was indigenous to districts of that type ; in fact, although the Leicesters, Merinos, and others at a remote period were so freely used at one time, it never became a typical longwool, and certainly is not now ; but in spite of its very altered type from when it was both the smallest and the finest woolled breed in these isles, it carries a splendid fleece. It is to be hoped that those interested in the breed will not sacrifice mutton features and quality to the endeavour of head covering and fat production, for, if so, some of its value for crossing will gradually disappear.

Primitive and Rarer Breeds.

There still exist several minor breeds of sheep, most of them associated with the islands from the Hebrides to the Isle of Man, which are, for the most part, of very primitive origin ; and there are a few others of which the origin is more obscure and probably of more or less recent origin. They do not play an important part in the sheep stock of these islands ; and as they have rarely undergone any careful selection or good management, they compare poorly with the breeds which are more commonly met with. However, as some of the sheep of note to-day were " weeds " in their unimproved condition a century and a-half ago, it is not at all impossible that under good management valuable sheep might be developed from them.

Visitors to the Bristol Show of the R.A.S.E. in 1913 had the advantage of seeing specimens of most of these breeds, together with sundry crosses, which have been gathered together by Mr. H. J. Elwes, of Colesborne Park, Gloucestershire. It was a most interesting collection, and Mr. Elwes is entitled to public thanks, as well as the appreciation of scientists and those interested in live stock, for the work he has been engaged upon for many years in connection with these types. No particular advantage would be served by such incomplete accounts of these breeds as could be given here ; and as Mr. Elwes has published a book on these sheep and the results he has obtained, together with numerous photographs, those who would follow up the subject would be well advised to get his book, which is published by Messrs. R. and R. Clark, Edinburgh.

Several of the breeds are of the polycerate type, showing the four horns ; and the more interesting breeds he deals with are the Manx, Soay, Shetland, Hebridean, Horned Black Welsh, Iceland, Papa Stour, Piebald, Siberian, and Fat-rumped.

THE MERINO.

The races dealt with in this book have been confined to those of British origin, consequently the Merino has not been discussed. Early endeavours to establish it in its pure state, or in cross with British breeds, were not regarded as successful, though there are a few flocks kept in the country which give their owners satisfaction. It has to be remembered that when the early crosses were made some century or more ago, British sheep were in a comparatively primitive state. It is by no means certain that, with the fuller knowledge of breeding possessed to-day and the wholly different types of sheep which have been developed meanwhile, crosses with British blood would not be more successful now; and it seems somewhat desirable that the question should not be altogether ignored. Very much more crossing was done in the early days than is commonly recognised, and there are records of crossings with most of the notable breeds of that time. The blood must have passed into ordinary stocks, and how far this made it easier for breeders to improve the fleeces of their several breeds is obscure; but considering the rapid improvements made it is far from improbable that the Merino gave considerable help. It is certainly difficult to see how otherwise the wool was so much improved in a short time, even admitting the very full use of the Leicester in early days; but the quality of the Leicester wool in the early days of the breed's improvement was never regarded as its strongest point.

CHAPTER VI.

HILL AND HEATH BREEDS.

CHEVIOT.

The Cheviot is a hill breed which originated in the hills of that name on the borders of England and Scotland. Everything points to the breed having long been established there, and a tradition places them as one of the breeds which the Spanish Armada contributed to these islands, but all these Armada traditions are of very doubtful credence. Still, the unimproved sheep was very distinct from the neighbouring heath black-faced breed, although the heath-type was undoubtedly there. How far the soil and pasturage influenced the colouring cannot be answered, but there is the instance of the black-faced, hornless Berkshire Knott having been produced alongside the old Wiltshire white-faced, horned breed. Originally a light-framed race, with short, close wool, it was brought into its more modern type by crosses with sheep from Lincolnshire; though whether of the old Lincoln breed or Bakewell's Leicesters is a debatable point; but Mr. Robson, of Belford, and two others selected sheep in Lincolnshire in 1756, and it is from crosses so derived that the present breed sprang. A white, or practically white-faced breed, hornless, with fine, clean wool, with mutton of good quality, a sound constitution and much virility has been formed. Originating on hill pasture of good, sweet herbage, the old stock has acquired a flesh-forming disposition of considerable aptitude, and it thrives best where the pasturage is full and good, though short, rather than on the heathy hills, where the black-faced sheep prove so valuable. Although the Cheviot Hills are the indigenous home of the breed, it is found pure, and often improved, on hill land of good pasturage over many parts of Scotland, even to Sutherlandshire; in fact, from the northern limits many of the sheep which prove so successful in the showyard and carcase competitions are derived. The light forequarters of the old breed have given way to a breed with a very striking forequarter. It is very impressive on other breeds, and the Cheviot character, once imparted, is permanently fixed. The full woolling of the neck, forming a ruff or frill, is very distinctive; it is noticeable in the Border Leicester, the Kentish or Romney Marsh, the more recently recast Welsh Mountain,

E

and the Exmoor breeds, and the quality of the meat of the results of the crosses improves, or remains excellent, although the flesh thickens.

A century or so ago the Cheviot made a great invasion into Scotland, and still holds much ground; but on the poorer and more heathy hills the black-faced breed is found more suitable. Sufficient regard was not always paid to the class of pasture on which the breed originated, and in such cases there was some degree of disappointment, and the black-face has, in many instances, regained its old haunts. On the better land it does well. In England it spread considerably from a century or so ago, and influenced permanently several breeds, but the coming of the Down breeds drove it back until it was comparatively little used in the southern half of the country. In recent years it has made a strong new invasion, and it is fast spreading in districts where farmers are not quite satisfied with the local breeds, and have to adapt themselves to market demands. It is a splendid race for crossing purposes, and, having taken a share in the recasting of so many breeds which stand well to-day, there is every likelihood that it will be called upon frequently in the near future to remould breeds to meet modern and future developments. The early dash of Lincoln or Leicester blood already mentioned doubtless helps the modern sheep to adapt itself so readily to so many and varied conditions, though it also probably accounts for its not being so popular on exposed, poor, heathy, hilly, or mountain land. Its breeding indicates good pasturage as being more desirable than heath.

The Cheviot crosses well with all Down breeds, and there seems to be a great future for it as smaller joints of highest quality meat increase in demand. At present the price of mutton is so high that size is being more appreciated than it was when mutton was difficult to sell; but as the world's supply gains ground, which may possibly take some time to do, there is little doubt, in view of the cold storage competition, that small meat will be in great demand. The increase in the number of Cheviots kept in the South of England is indicative of this.

The Cheviot lambs rather late in its native district, beginning about April 16. The lambing generally takes place on the hills, only those needing assistance being brought into the fields. The sheep are generally sold to feeders to be fattened on turnips. Wedders, sold off turnips at about ten to twelve months, weigh from 50 lb. to 60 lb. The hoggs cut about 5 lb. of wool. A shepherd looks after 600 to 700 sheep on the hills, having extra help at lambing. The sheep get no hand feeding except during a storm, when from ½ lb. to 1 lb. hay is given daily whilst the snow lies. A few flockmasters give a little artificial food in spring, in the shape of maize spread on bare pasture, but this practice does not develop, as ewes are generally found to mend better when they have had no extra assistance.

The following is a good description of the Cheviot sheep :—

A Cheviot tup, when arrived at maturity, weighs, when fat, at least 200 lb. live weight. He should have a lively carriage, bright eyes, and plenty of action. His head should be of medium length, broad between the eyes, well covered with short, fine white hair; his ears, nicely rounded and not too long, should rise erect from the head—low set, or dropping ones, are a decided fault, but at the same time they should not be what are called "hare lugged," that is, too near each other, as that indicates a narrow face, which generally denotes a narrow body. His nose and nostrils must be black, full, and wide open; his neck strong, and not too long; his breast broad and open, with the legs set well apart. His ribs must be well sprung, and carried well back towards the hook bones, as a long, weak back is about the worst fault a Cheviot can have. His back must be broad and well covered with mutton; his hind quarters full, straight, and square; the tail well hung and nicely fringed with wool. His legs must stand squarely from the body (if bent hocks, either out or in, the latter especially, are looked upon as a weakness); the bone must be broad and flat, and all must be covered with short, hard white hair. He ought to grow a fleece weighing 10lb. or 12lb. of fairly fine wool, densely grown, and of equal quality, coarseness on the top of the hooks is a decided blemish; the wool should meet the hair at the ears and cheeks in a decided ruffle; bareness there or at the throat is inadmissible, and it should grow nicely down to the hocks and knees. The belly and breast ought also to be well covered.

The same description, when modified, will apply to ewes also, which will weigh 100lb. to 150lb. Cheviots, when in a natural state, must grow finer wool, as hard feeding inclines to make it stronger, but it must be stiff and dense, and not too short. A hill flock should clip on the average 4¼lb. each; if wedders are kept that average will be increased. A lot of draft ewes, when fed moderately fat, will weigh from 95lb. to 100lb., and old wedders 160lb.—live weight. It must always be remembered that along with feeding qualities Cheviots must embody great hardihood and milking properties, for they are expected to stand great privations on their native hills in hard winters and backward springs; for if once they are bred too soft to live and bring lambs on their own ground, they either die off, and the ewes cannot be kept up, or their owners have to incur great expense in taking grass parks or other keep, which, with wool at its present price, is a ruinous business. The perfect Cheviot is one which will live and thrive well on the hardest keep, and when taken to lower and better ground, prove itself equal to the occasion by growing larger and becoming fat.

EXMOOR OR PORLOCK.

The Exmoor is one of the four breeds associated with the south-western part of England. It is indigenous to the Exmoor Hills in West Somerset and North Devon, and was a heath or forest breed, small, of almost goat-like appearance, possessing great agility. Low described it, about eighty years ago, as being much crossed out by the Cheviot, and doubtless the splendid sheep which has been developed owes much to the Cheviot, the true Cheviot frill of wool around the neck and throat clearly indicating this, as does the general build, which contrasts greatly with the original form. But the hardy constitution remains in this well-horned breed, which retains in so many ways its old heath character— in spite of its increased size—and good fleece of close set, well stapled wool. Moreover, a slight slackness behind the shoulder, noticeable in the shorn sheep, further suggests this. It is not merely as a local breed that the modern Exmoor proves its value, for it is one of the most prominent of the minor breeds which are being largely taken up in different parts of England. It is hardy and a good thriver; the ewes are good mothers, and will do well on poor land, though, as for a long time they have been accustomed to receive roots, they do well in the fold, as well as on high exposed grazings. The fleece weighs about 5 lb. The Exmoor is as attractive in appearance as it is useful, and breeders can confidently look for an extended demand for this little breed, which, with high-class mutton, eminently meets modern demands.

DARTMOOR.

The Dartmoor, now a long-wool breed, has probably undergone as much alteration from its original type as any breed in these islands, for it was at one time very small, much like the sheep of Wales, carrying a long, soft wool, of different type to the ordinary heath or forest races. It was then whitefaced, and the males carried horns. They, however, mated extremely well with the Leicester and Southdowns, though the Longwools made most impression on them. The original Okehampton mutton, much favoured in London, was produced by the Dartmoor and Exmoor sheep. Now the Dartmoor sheep, as wethers of one and a-half to two years old, weigh from 20 lb. to 24 lb. per quarter. The horns have been bred out, except for an occasional snag on the rams. The wool, growing on a clear pink skin, is often 15 in. in length, of very fine quality, and ewes yield up to 10 lb. or 12 lb. The sheep is well covered at all parts, and the meat of superior quality to that of the essentially long-wool breeds.

It is a distinctive breed in which the forest and long-wool traits, with a slight Down trace, from which the facial colouration is derived, are all distinctively marked, though well amalgamated. Dun markings which, historically, are a relic of the Welsh character (Wales once included Cornwall), are not approved, but black spots of hair on black skin, deepening in the neighbourhood of the nose and round the eyes, are regarded as characteristic. The sheep are economically kept mainly on grass with a little hay, and the ewes, which are good mothers, cross well with the Down breeds to produce fat lambs. By good selection the breed has been enormously improved, without impairing its usefulness on the exposed pasturage which it grazes.

WESTERN OR OLD WILTS BREED.

The Western or Old Wilts breed became extinct in its native district some eighty years ago, owing to its being crossed with other breeds, notably the Southdown, when it became the foundation stock of the Hampshire Down. A small offshoot of this breed apparently was taken into Wales, where it missed the Down crossing, and has remained fairly true to type, though improved by selection. The sheep are valued for crossing with draft Mountain sheep to produce fat lambs, and in parts of Bucks and Northamptonshire small flocks are seen. They are prized for fat lamb production. These districts were much visited by cattle and sheep dealers from Wales for many generations, and doubtless in this way they reached the farmers in the South-East Midlands, who use them on any breed for lamb purposes.

There is difficulty in maintaining the breed in a pure condition owing to the small numbers remaining, and in some cases the Dorset Horn is being used on the old Wilts ewes. The quality of the meat is good, and their popularity for fat lamb indicates good thriving properties. As a wool sheep they have little to recommend them as the fleece is light, and they share with some of the older unimproved heath breeds, though to a greater degree, the characteristic of peeling or shedding their fleece. With the older sheep this takes place in April or May, and with the lambs in July. There is no Flock Book. It is an old heath breed with strong horns, something of the type of the Dorset Horn, but they spring more from the crown and throw out rather wider. Practically a whitefaced breed, they are greyish about the muzzle, and the pelt carries dark spots.

Welsh Mountain.

The native sheep of Wales are of the mountain type. There are two original stocks—the one of the higher hills, a naturally wild, dark-faced, horned sheep, and the other a lighter faced, white or tan, the soft-woolled, pink-nosed, antelope-looking breed, also horned, of the lower hills. There are many variations of these, as is natural in country where there is much alternation of hill and valley, and of deeper or shallower soil, as well as of geological formation. The breeds are of great antiquity, and, as such, have acquired indigenous character that adapts them to their native hills. However, as they meet with more liberal treatment owing to agricultural development, which was greatly furthered by the Enclosures Act, they have been altered to suit modern conditions, and a steady improvement is being maintained, especially with those that generally affect the moderate altitudes. The quality of the mutton has always ranked very high, even in the distant London market.

In the early days of sheep improvement the usual endeavour to improve the sheep, especially those living on the lower levels, was made, although it was not pushed as far as with most breeds. The influence of extraneous blood of other mountain breeds, such as the Cheviot and the Black-face in the early days, was not great. The improvement of the several Welsh breeds and variations was brought about more particularly through the Shropshire, the Clun Forest, the Cheviot (at a later stage), and the Ryeland, but the Ryeland's influence was best infused by the blood of that breed that existed in the Clun. The Scotch Black-face has been tried, but with variable success. Over many parts of Wales, on a better pasturage, the Shropshire pure, but more often in the cross, is met with in considerable numbers, and in some counties the Ryeland is kept pure. The existing native Welsh breeds are the Welsh Mountain, the Kerry Hill, the Radnor, and the now scarce Plinlimmon; the Clun originated in Shropshire.

But the Welsh Mountain Sheep, as now recognised by the Flock Book Society, founded in 1905, has definite features and characteristics set out to distinguish it from the other breeds. Whatever other races may have been infused into it during the period of chance breeding that preceeded straight breeding of more recent years, there is practically nothing to indicate any except the Cheviot, and that in only a moderate degree. In those of the breed that has been markedly handled by the breed-maker, the antelope neck and light fore-quarters have given place to an altogether fuller fore-quarter. The carcase, which is somewhat long and narrow, carries a fleece of short, fine, thick wool, and occasionally flocks of black-woolled sheep of high quality, descendants of very old strains, are met with.

The rams have strong curved horns, and the ewes are hornless; the face and legs are white or tan, the latter being preferred where the sheep are required to go on to high pasturage, as they indicate hardness of constitution. A kempiness or breechiness in the fleece is not objectionable for the same reason. The lambs at birth usually have a well-marked dirty yellow patch on the back of the neck, which shortly disappears. A sheep has been developed which is suitable to be brought out earlier, and, with the arable land having been brought more into sheep farming, many wethers are fattened out on roots in their first year, a great contrast to the three or four years which the unimproved sheep required.

These sheep now represent a high-class mutton breed, with small joints. The sheep naturally kept weighing about 30lb. dead-weight, though, on rich low ground, where they are well treated, they may reach 50lb. upwards. The fleece cuts about 2½lb. for ewes, and 3lb. to 5lb. for rams.

THE KERRY HILL.

The Kerry Hill is a Welsh breed, taking its name from a big parish in Montgomeryshire called Kerry. A Flock Book was published for the first time in 1899. They were one of the many strains of Welsh sheep, and their improvement was due to the persistent energy of the farmers about Kerry, and to the good pasturage they found. Mr. Holford, who did so much to further the interests of the race, remembered them seventy or more years ago as a sheep with an all-white or slightly spotted face, with legs to match, and generally of a good type for Welsh sheep of that time. Up to about 1840 to 1850, when much of the land was enclosed, farmers relied on their own stock. He further expressed to me, that at the earlier date it would be fair to say that the Kerry was an intermediate between the Clun and the Radnor until the Shropshire became a noted breed. With the better breeding opportunities enclosing afforded, farmers went further afield, and the Kerry men went to Knighton for stock rams, where they got the Clun Forest sheep at the time, a more advanced breed, raised on good land facing the Kerry Hills. The sheep rapidly improved from the Clun, and also from a slight dash of the Shropshire, generally through the Clun.

In course of time the Kerry men had brought their sheep to type, and selected them to the white with black spots, eliminating the tan, which was characteristic of the Clun. In the main, for the past forty years they have largely depended upon selection from among local flocks, so as to keep the character. The Clun, in the opinion of some leading breeders, has, as years passed, been more influenced by the Shropshire than they thought desirable. Nevertheless, for a time a good deal of interchange went on, and, some twenty years ago, bar the leg and face markings, there was considerable similarity in flocks described respectively as Kerry Hill or Clun Forest, whilst some of the latter showed more Shropshire influence.

The amalgamation of breeds, and subsequent selection, have produced a remarkably vigorous and hardy breed, not strictly of mountain type, although possessing many of its features. At present the horns are not quite bred out of the rams, though nearly so. It is an excellent grazing sheep, but is well suited to winter on roots, and generally to be kept on arable land adjoining hills, where it can have the run of both. It produces a good carcase of first-class meat, with a fleece of 4lb. to 8lb. from ewes and wethers. The wool is somewhat kempy, and where it is required to exist in wet districts, it would not be advisable to get rid of this, it ensures the rentention of mountain characteristics, and is in keeping with the distinctive long, large and fleshy tail. The breed has become popular on heavy and wet grazings in many parts of the country, where it is met with either pure or crossed. It has proved itself a good sheep on parts of the Sussex Weald. The ewes make good mothers, and are in much demand for crossing with the Welsh Mountain, Shropshire and other breeds to produce early lamb. It is a breed that appears to have great possibilities in countries abroad where there is hilly, cold, exposed, or wet land.

Whilst the Kerry Hill Sheep Association publishes a standard of points which it thinks well to place before breeders for general guidance, it recognises that there is yet some diversity of type, as is only natural in a breed which has not been a long time under the guidance of an association, and evidently desires that latitude should be given to breeders to mould the type as experience best guides. Among the breed characteristics which they put forward are :—

Head.—Well covered, but brown and black on top between the ears is objectionable.

Face.—A good speckle, black and white, not too black. Clean cheeks, well woolled to the jaw bone.

Ears.—Fairly short, thick, well set, speckled.

Tail.—Long, well set on fleshy dock, with plenty of wool to the point.

Legs.—Large bone, speckled, free from wool below the knees.

Skin.—A nice pink or red skin, free from black or blue spots.

Wool.—A tight, close fleece of white wool, a good length, showing a little fledge on the face, coarser on breech and tail. Horns objectionable.

CLUN FOREST.

The Clun Forest breed was long established in the forest of that name, which represents a tract of some 12,000 acres of hills of considerable height, with accompanying dales of good feeding. Very much of the land has been enclosed for many years, and a large portion of it has been brought under the plough. It is situated in the south-west corner of Shropshire; therefore the breed does not rank as a Welsh one, although it has considerably influenced the neighbouring breeds over the border. The Clun Forest joins the Kerry Hills, but before the Enclosures Act, when the sheep ran alongside one another, they grazed apart and rarely intermixed. The Clun Forest breed was an offshoot from the Old Ryeland, which had long existed on the richer land about Clun, and acquired distinctive features; it possessed good thriving properties, and no mutton was held in such high estimation throughout many years in the West End of London, where the best butchers always announced they were purveyors of Clun mutton, which, at that time, was small and excellently flavoured.

The Clun played a great part in the making of the Kerry Hill breed, but with more land locally coming under the plough, the Shropshire was crossed on it very freely, and some of its characteristics were altered. It is still a splendid sheep, and it seems regrettable that those interested in it have not taken steps to preserve it as a distinct breed, for few sheep have better inherent features. However, some of these are preserved in the Kerry Hill race; and, with this breed near by, as representing a type of the hills and adjoining arable land, and with such a powerful high-class breed as the Shropshire dominating so much country on the other sides, it may be difficult to find a place for it in a native area which is not large. It is, however, a type of sheep that ought not to be altogether lost to a country. It is now fast merging into the Shropshire type, but the forest characteristics are by no means eliminated. It cuts a less kempy fleece than the Kerry. There is reason why, at some distant time, should the Shropshire in some localities require a breeding back to a more original stock, that some mild infusion of the Clun should act beneficially.

THE RADNOR.

The Radnor breed was held by Low to have much in common with the sheep of the higher Welsh mountains, which lost much of its excessive natural wildness as it occupied lower ground, where it increased in size, though it has not acquired the size of the Kerry Hill. But few of the original Old Radnors remain; crossing with the Kerry Hill and the Shropshire having been very generally adopted. The west of Shropshire, and the adjoining Welsh counties, used to possess a great number of local breeds, many of which were overpowered by the Shropshire, as was only natural where so excellent a breed, with great adaptive powers, was so readily available. Even yet the possibilities of this big district for breed making are not yet exhausted, and it must always be an interesting locality for the sheep breeder. The Radnor itself was not as good a stock to improve from as were some of its neighbours, and it has been comparatively left behind and overshadowed.

THE BLACK-FACED MOUNTAIN.

The importance of this breed is proved by the great number of sheep entered in the 1908 Census as belonging to it. It is widely and numerously found in Scotland, and is often spoken of as the Scotch Black-face breed, as Scottish breeders have especially identified themselves with its improvement. It is of the heath type, though, through many centuries of existence on high hills or mountains, it has developed very fully the characteristics of mountain sheep, and it is as a mountain sheep that it is regarded ; yet, when brought to the lowlands of England, the sheep show their ready adaptability to lowland conditions ; in fact, in recent years they have made a great southern invasion. Moreover, they may be met, thriving well, in parts of Ireland, which are of distinctly lowland type. But, on the hills of Wales, they do not make their mark so deep, the Cheviot being recognised as being more suitable, and as making a better cross, with the Welsh hill sheep. This does not imply that the Black-face cannot be kept profitably there ; but that the conditions more favour the local and the Cheviot breeds. Through the north of England, where the land is largely of a hilly or mountainous character, it holds sway. In the Kerry Mountains of Ireland it has practically crossed out the native short-tailed breed ; it has crossed out, or substituted, the old native heath sheep or cottagh ; in fact, its range of useful habitation is very wide indeed. It not only thrives well over this widely spread area, but the high quality of its meat secures it a good market at all times. Were the narrow origin which some would ascribe to it, especially that of those who emphasise the idea that it traced from the old Scotch dun-faced sheep, it would be very unlikely that it would possess the ready aptitude to acclimatise in so many varying conditions, or that it would mate so readily with other heath breeds.

It has much in common with the dark-faced heath breeds which originated the Southdown, the Suffolk, and others found further north, until they sought the higher hills, and gradually extended their way through Scotland; though, as Low pointed out some eighty years ago, it was not until the middle of the eighteenth century, when black cattle gave way to sheep, that it was met with in Argyllshire and the central and northern Highlands. It had, however, existed in Dumfries, Berwick, Roxburgh, Selkirk, Peebles, Lanark, and adjoining districts, for an unknown period. The main feature, however, is that it has the characteristics of the heath breed, which is responsible for the finest mutton found in these islands, or, for that matter, in the world, and beyond this the adaptability to repopulate, almost without regard to altitude, the districts which indigenously carried heath breeds before relatively modern introductions pushed them aside. The Black-face is fast helping to fill in these lapses, especially where the sheep recently kept have not been well suited to the soil and conditions generally. As it crosses so well with many breeds, and as breeds in the cross with it maintain a high quality of meat, suited to modern needs, it is a breed that must almost certainly be used in future to remould old, or substitute new, breeds. Although it thrives under conditions which few sheep could withstand, it prospers with well-doing.

The "Cross" sheep of Scotland are almost exclusively of the Black-face or Cheviot with Border Leicester. Wensleydale rams are frequently used in the south-west of Scotland and the north of England, known in Scotland as Yorkshire crosses, and in England, especially in the south, to which they are freely sent, as Mashams. Many of these come far south, and not a few are turned into the London public parks for grazing.

As the breed is found over such a very wide area, embracing many soils and climates, there is naturally some variation in type ; but, on the whole, this is well maintained.

Scotch Black-faced mutton is highly prized wherever well-flavoured meat is in demand ; that from old sheep, from three to five-year-old wethers, weighing 15lb. to 16lb. per quarter, used to have preference, but taste has changed, and younger has preference, particularly wether lambs of about 36lb. The average fleece is 3½lb. to 4½lb. for ewes, and 7lb. for wethers.

The accepted points of the modern breed are :—The face and legs are black or mottled (distinctly clean, and free from dun or brown), smooth, and glossy. Dark faces are in favour with breeders of cross-lambs, which bring higher prices when dark. Wool should not appear among the hair, although a slight tassel on the forehead and fringes on the cheeks and legs may have to be dressed off well-bred sheep for showing. The nose is strong, broad, and prominent, and the nostrils are wide and black. The horns of the ram are large, coming out level from the crown and taking one or more spiral turns, according to age. The short ears are hidden by the horns of the ram. The tail is not docked, being naturally short—reaching only to the hocks.

A more detailed description of the management of the breed is given in a later section, " The Management of a Hill Flock."

LONK.

As in all other parts of England before the country was well opened up, and until, from a century to a century-and-a-half ago, sheep owners set out to improve their sheep, there were many local breeds in the northern counties, from Derbyshire upwards; many had characteristics in common, accordingly as they were indigenous to the richer lowland pastures, or the sparser grazings of the hills and woodlands. In the hilly districts, the geological formations, the elevations, and climate, had considerable influence in the moulding of the features, so that while some approached somewhat nearly to one type, others moved very considerably from it. They, however, partook of the old heath type, whether they were black-faced, white-faced, horned, or hornless. With greater care in sheep management there was more selection, and gradually the inferior were pushed aside, until comparatively few were left as representative breeds. Among the more important is the Lonk, a native of the wet hill districts of East Lancashire, West and South-west Yorkshire, and North-west Derbyshire. It is of the type of the Black-face, but is longer in the leg, and longer and bigger in the body and head. It is horned and has a black face with clear white markings; it has a closer, heavier, and finer fleece than the Scotch sheep, but is not quite so hardy. These are features which would be expected from its environment, and from the better class of herbage that it is indigenous to. It is evidently indigenous to the country it affects, its name being a curtailment of the old provincial pronunciation of Lancashire —Lonkashire. It is one of the breeds which more distinctly links up the lowland heath breeds with the heath breeds which have long worked their way to the higher and colder hills, and have acquired more truly mountain habits and characteristics, as shown in the black-faced sheep of Westmorland, Cumberland, and the Scotch hills. When the Lonk breed was first recognised as such to distinguish it from allied or neighbouring breeds is not clear, and there is nothing but quite modern reference to it by that title. Coming from relatively good pasturage, the sheep has many valuable characteristics, and in crossing it mates well with such high-class mutton and wool breeds as the Wensleydale, Leicester, Hampshire and Oxford. It is one of the most recent breeds to aspire to a flock book—that phase having been reached in 1905.

Washed fleeces of good wether Lonk hoggs run from 7lb. to 9lb., and of good " gimmer " hoggs from 6lb. to 7lb., and ewes 4½lb. to 6lb.   Mr. W. Ralph Peel gives the following very lucid description of the points of the breed in the Journal of the R.A.S.E. :—

The head, which is very characteristic, should be black, or white, or black and white.   The favourite face colour is black, with distinct white markings, brown or mottled faces being objected to.   These markings, together with great depth and strength of lower jaw or chap and a pronounced " Roman nose "—points strongly insisted upon by breeders—give a handsome and picturesque appearance, perhaps unrivalled among all our breeds of farm stock.   This fine head, which should show width between the eyes and over the nostrils, must carry massive and well-placed horns.   Great attention should be paid to how the horns are placed on the skull.   They should be set on wide apart at the base, should come from the skull nearly level with the top of the head, and have curl enough not to meet the cheek.   The breeder believes that a strong horn showing quality—that is, a clean horn with not too many puckers on it—denotes that strength of constitution without which this favourite sheep is unable to battle against the many adverse conditions found among his native hills.   The horns curl once or twice according to age.   The head and horn should be well set off by an eye standing well out, showing a fine and bright colour.   Width of neck and loin are insisted upon, and, as far as is possible, narrowness of the shoulder is to be avoided.   If this defect, characteristic of hill sheep, cannot be altogether eliminated, depth through the heart and well sprung ribs must always be typical features in the frame of the well-bred Lonk.   The tail, which, in the ram, is always undocked, must be strong, wide, and long ; in many well-bred lambs it not only reaches, but drags along, the ground.   The legs, which should be black and white in colour (too much white being often looked upon as an indication of softness), should be well formed, strong, and showing ample bone.   The wool, which should come well up to the horns, and down to the knees and hocks, is exceedingly fine in texture, close, or thickly set, with no flake or curl.   The wool, when washed, has a very fine silky appearance, is perfectly white, and is very valuable in the manufacture of various fine goods.

Derbyshire Gritstone.

The Gritstone or Dale-o'-Goyt breed has been long established in the hills and dales of the Peak District of Derbyshire, where, for more than a century, flocks have been kept pure.   It is of the heath type, but, unlike most of the heath breeds, it is hornless. It has a dark face, preferably black, with white mottling, and the legs partake the same colouring.   It was only in 1906 that the breeders formed a society to forward the breed's interests. They were well warranted in doing this, for an undoubtedly good sheep has so been preserved.

It has little of the appearance of a native breed recently taken in hand, for it shows up well beside many of the older established breeds. It is capable of producing a good carcase of high-class mutton, has a well-balanced frame, a vigorous constitution, and cuts a good fleece of wool of what may be termed the finest type. The breeders will be well advised to continue on the lines which the best men in the breed have followed so long—to pay special attention to the mutton, whilst maintaining the wool of a quality and nature which suits the climatic conditions of the bleak hills with which the breed is associated. At the same time there seems no reason why this breed should not be mated successfully with several other breeds, not necessarily hill sheep, to produce cross-bred mutton of high quality.

SWALEDALE, PENISTONE, AND OTHER ENGLISH HILL BREEDS.

Among the remaining breeds of the northern hills may be mentioned the Swaledale, a very hardy breed, found on the hills adjoining the dale, and extending into the Pennines and Westmorland. It is much of the type from which the Scotch Blackface was developed. In some respects the sheep might be regarded as superior to the Scotch, but they are excellent mothers, and live a hard life, getting no extra food beyond a little hay in severe weather in spring.

The Penistone is now a small breed met with occasionally on the borders of Yorkshire, Lancashire, and Derbyshire. It is heavily horned, and has a white or light-grey face, with something of the Lonk characteristics.

The Limestone has practically become extinct. It was chiefly found in a few parishes on the dry limestones of the lower districts of Westmorland. In a district, and over a large breadth of country, where most of the sheep were black-faced, it was singular that, although horned, the face, and legs, and wool were white, again illustrating how wide could be the variation in breed characteristic within small areas.

The Rough Fell sheep of the moors and hills of North-west Yorkshire, parts of Westmorland, and adjoining districts are closely allied to the Scotch Black-face, but have not had the same skill and attention generally bestowed on them.

WICKLOW.

The Wicklow is an Irish breed with many points in common with the Welsh Mountain. Moreover, it possessed, in its unimproved state, quite as good features as those of the better Welsh breeds. Ireland, however, did not march with Great Britain in the general improvement of a century or so ago, and the Wicklow sheep was not taken in hand by the breed improver. The breed has been much cross-bred in the past quarter of a century, and pure strains are difficult to find. Still there are sufficient sheep showing many of the natural features, from which a good breed might be developed. That a good race, well suited to the requirements of a considerable portion of the south-east of Ireland could be built up, there is little doubt. It is distinct from the old Irish Cottagh breed found until within recent years in some parts of the West of Ireland.

THE KERRY MOUNTAIN SHEEP.

The Kerry Mountain Sheep is an Irish breed associated with county Kerry, but in the past quarter of a century it has been greatly influenced by the Scotch Black-faced sheep, which has mated well with it, and the cross suits the country. Originally it was a short-tailed breed, with many goat-like attributes; doubtless of singular origin to the oldest breeds known to have inhabited these islands; but now only met with in the extreme north. A few years ago, I spent a considerable time on the Kerry Hills trying to discover a flock of this breed, or even one only a little crossed, but could not find one, though twenty-five years ago they were common enough. To all intents it is a breed of the past, and the crosses showing much of the Black-face character, mainly represent the sheep at the Kerry fairs.

HERDWICK.

In point of numbers the Herdwick ranks highest of the northern dark-faced breeds, excepting the true Black-face. Its origin, traditionally, is another instance of the custom of attributing the uncertain to the Spanish Armada. Whether they came in this way, or are merely modifications of hill sheep of the north which emanated as a breed kept in a small area about Muncaster, is open to question, but that they ousted a white-faced horned breed, an indigenous sheep known as the Old Fell breed, is fairly certain. The old breed was larger, but it was slower to feed, and less profitable to keep. The Herdwick holds its place, and is likely to, because no other breed has yet succeeded in turning to such good account the short herbage on the Fells. One is much inclined to sympathise with another writer sceptical to the imputed origin, when he said, " If they were the victims of a shipwreck, they picked their country well." The breed exists under some difficulties owing to the large amount of commonage where the rights are exercised to the full and more, so that the spare ground is often too severely taxed.

Although the legs and faces of lambs are black, or black with white flecks, by the time they are two years old they have become a frosty or silver-grey, with the forehead a little darker than the lower face. But there is a distinctive blue-black mark or patch at the back of the neck. The ears are white and sharp, and the whole demeanour suggests the wonderful activity that allows them to scale apparently inaccessible places, and to jump both high and wide. A custom prevails of holding a stock to its particular ground or leaf, to which it is much attached, and the tenant's lease usually has a clause which gives the landlord right to the refusal on outgoing. The breed may be regarded as almost invaluable in the country that it affects.

The distinguishing characteristics of Herdwick sheep are :— Horns in the ram (the ewes have no horns) should be smooth and not too thick, coming out of the head well apart and well back ; the face a light " rag " (grey) or white in full-grown sheep, with plenty of white bristles on the back of the head, and a " toppen " of moderate size on the forehead. The neck and head should be carried gaily, rising well from the shoulders, which are usually sharp at the withers, although a broad shoulder is preferable. The fleece should be genuine wool, not hair, the staple strong, with a mane standing well up round the shoulders and down the breast ; the wool a good length on other parts, and knit together with a lash on the top. A little kemp in the wool when a sheep gets to six years old indicates true Herdwick character. The animal should walk freely and be square on his limbs in travelling to and from an observer, and have a good thick tail. The best sheep, when turned up, are grey below, and they are none the worse for being grey all over the body—showing a grey pelt after shearing. Ewes clip about 3lb. of wool, hoggs and wethers 4lb. or more, and rams 7lb. or 8lb. Ewes do best when they lamb for the first time at three years old. Hoggs from some farms are wintered in the low country at a cost of 5s. to 5s. 6d. each, but on very high exposed places they are hardier and do better afterwards if wintered at home on hay.

CHAPTER VII.

LONGWOOLS.

THE LEICESTER.

The Leicester is the great historic breed of this country—from more than one aspect. It was the one to which modern methods of selection and mating to establish a new breed were first applied, and it exercised an influence over almost all British types at a time when they were in a very crude and uneconomical condition. It did this in comparatively few years, with a success that revolutionised the sheep stock of this country, and, to a great extent, throughout the wool-growing world. Moreover, although as a breed its direct influence and importance have diminished, the breeds which came under its earlier influence still retain in a prominent degree, and as essential to their success, characteristics and features imparted fully a century ago. It was on a local breed, one of the long-wool type of the lowland of the Midland Counties, but not marsh pastures, that Bakewell worked, when he started at Dishley, near Loughborough, to improve sheep—about 1755. Since then there have been many breed makers, but his was the master-mind who, with nothing of an historic experience or knowledge to guide him, created a new art or science, which has been almost invaluable to the world, and has been applied to practically all domestic animals with such marked success. His work was very quickly crowned with success; he obtained a fixed type which could be imparted to the offspring, secured earlier maturity, and, by selection, moulded the sheep into type to which all mutton-making breeds have been assimilated; he paid less attention to the wool, but as the wool of almost all breeds was improved by the Leicester cross, he did not leave much to be desired. In giving special attention to the mutton features, and thus educating sheep-breeders what to aim at, and how to attain it, he was most usefully employed.

It was somewhat remarkable that Bakewell should have had at hand a breed that was so admirably adapted to be dealt with as a pioneer, and at any rate it showed his great discernment in choosing it. The propensity to fatten, and to attain early maturity which the improved Leicester developed, would probably not have been attained in so marked a degree had any other race been taken in hand similarly. Its wonderful heart girth was an asset of greatest value, as it is in all meat-making animals. In course of time some of the breeds which had been brought under the influence of the Leicester made very strong rivals. The tendency to produce fat became excessive in the light of popular taste. The Leicester was somewhat lacking in constitution, and generally proved more

F

valuable in the cross than pure, when moved from its native district, though, strangely, it did better as it went northward than when it was taken to districts southwards. Its relative—in comparison with other long-wools—produce of wool told against it at a time when wool was very dear—and gradually the number of pure flocks decreased. In fact, so far as its native district and the counties around it were concerned, the flocks became very few, though further north, as in Yorkshire, Cumberland, Westmorland and Lancashire, interest in it was better maintained, possibly because the operatives in the big industries of these counties maintained a partiality for fat meat, whilst those in other districts lost it. There seemed almost a probability of the breed flickering out so far as the home trade was concerned, but in 1893 the Leicester Sheep Breeders' Association was formed and, owing to their exertions, new life has been instilled into it. A much stronger type of sheep has been built up, and the fleece has been greatly improved. In fact, at no time in its career has the breed shown such a vigorous type, and it retains its old characteristics for early maturity and the laying on of meat. How well the modern breeders have recast and rejuvenated the breed received full demonstration when Mr. Jordan was awarded, in 1907, the Championship in the Longwool classes at the Smithfield Show.

The demand for cross-breeding purposes is considerable, for though primarily a grass sheep, it has throughout more than a century been used pure or in the cross for winter feeding on roots. Its quiet disposition make it suitable for the fold ; but in the great root-feeding districts, sheep of the Down type are generally more favoured, especially as winter folding on roots is an expensive system, and high-class mutton is necessary to meet the expense. Where grass is more relied upon, the Leicester cross is more favoured. There is an offshoot from the breed known as the Border Leicester, which differ considerably in several features from the Improved Leicester. The term "Improved" Leicester, by which name the breed was known in early days, to distinguish it from the unimproved, has long been considered redundant, though at one time it was a name to charm with. In its present type it is likely to gain in popularity for crossing purposes and for regenerating stock abroad.

The "points" of the present-day Leicester may be summed up as follows : Lips and nostrils black, nose slightly narrow and Roman, but the general form of the face, wedge-shaped, and covered with short white hairs ; forehead covered with wool ; no vestige of horns ; blue ears (sometimes white), thin, long and mobile, a black speck on face and ears not uncommon ; a good eye ; neck short and level with back, thick and tapering from skull to shoulder and bosom ; breast deep, wide and prominent ; shoulders somewhat upright and wide over the tops ; great thickness from blade to blade, or through the heart ; well filled up behind the shoulders, giving a great girth ; well sprung ribs, wide loins, level hips, straight and long quarters ; tail well set on, good legs of mutton, great depth of carcase, fine bone, a fine curly lustrous fleece (the sheep are well woolled all over) free from black hairs, with firm flesh, springy pelt, and pink skin. The general form of the carcase is square or rectangular ; legs well set on, straight hocks, good pasterns, and neat feet.

BORDER LEICESTER.

The Border Leicester is a breed in which the Bakewell-Leicester is paramount ; many claim that it is purely descended from this improved Leicester, and, on the showing of such records as are available, there is good support of this ; but as the earliest of the Bakewell-Leicesters were taken into the Border districts by Mr. Culley in 1767, and others soon followed, it would not be difficult for some little unrecognised uncertainty to have crept in, because, until flock books were established, records were kept in an amateurish manner. Moreover, there is no obligation to abide by direct breeding in those circumstances, and, for that matter, he is a bold man who would assert that in the past quarter of a century no foreign blood had been brought into some of the well-known breeds. In the early days of sheep improvement, which was largely instigated by the success of Bakewell's Leicester on other breeds, there was a vast amount of experimental crossing —and it would be strange if this improved sheep should not have been used on the Cheviot. But all must be surmise where actual knowledge is not available. If, however, the Border Leicester had no Cheviot in it, the change from the Bakewell-Leicester, as effected by change of environment, must be the most remarkable among our breeds. In fact, if there was not a slight dash of Cheviot blood strained into the border-kept Leicester in remote days, all one can say is the two breeds must have looked long and longingly at one another until some of the elements of sympathy which are said to influence yeaning animals wrought changes. But this is not likely. The change is so pronounced that the Border Leicester breeders moved for separation from the English Leicesters, and there is, indeed, great distinctiveness between them. Some day breeders will more fully recognise the advantage a breed possesses where, in its formation at a remote period, more than one breed took part, and as there is no breed in which it can be proved there has not been a remote admixture, there is little advantage in urging that a breed is poor in the numbers of factors which went to its composition. Wider usefulness and adaptability for wider conditions are associated with the more composite breed, a breed with only one set of indigenous characteristics has narrow adaptability.

The Border Leicester breed contributes largely to the success of the farming of the Border district. It is largely employed to cross with the Cheviots and Black-faced races, and the offspring of these is popular in parts of England; in fact, the crosses work well on pasture or roots, producing good weights and maturing early. Under the cross a considerable number of fat lambs is raised. These crosses come south in very considerable numbers.

The following are characteristics of the Border Leicester ram :— The head is long, and well carried on a neck of good length and ample substance at the base; broad, but not high on the crown, nor too heavy behind the ears, the two latter points in the ram involving difficulty of lambing in this as in other large breeds. Too much strength in the head is frequently correlated with coarseness in the animal. The profile should be slightly aquiline, with a strong masculine appearance, tapering to a black and square muzzle; the dense covering of hair on the face and legs uniformly white and hard (but not so wiry as in the case of the Cheviot), free from any trace of wool, and extending well back behind the ears; the ears, fairly erect and a good size, placed not too wide apart, white inside and out, and occasional black spots; the belly light, carrying little offal, and giving a somewhat leggy appearance, especially after shearing; the wool long and close, soft, and in little ringlets or pirls, wavy throughout its length, but not open to the skin; on being gripped it should fill the hand; the ram should carry a heavy fleece with the wool well down on the legs and with the belly well covered. Sheep deficient in the latter respect are not so well fitted to withstand unfavourable conditions.

THE LINCOLN.

The Lincoln breed is the typical white-faced, hornless, long-woolled breed, and carries a long loose forelock. Its foundation was a very big, coarse, heavily-woolled sheep of a type found in rich fens and marshes, and, as such, was distinct from the long-wool sheep of more upland pastures, which were commonly met with in Leicester and contiguous counties. When the improvement of the Leicesters was being recognised, the Lincoln men generally tried to improve their breed by selection, and achieved fair results; but wiser counsels prevailed, and the Leicester was brought in, and from repeated crosses with the Leicester the modern Lincoln was evolved. Its great size and coarseness gave way to a smaller (though by no means small) sheep; neater, and possessed of much more rapid maturing powers. The sheep still run to great weights, and work well on pasture, or on " seeds," rape, and other root crops. Lincoln mutton does not appeal to those who have been accustomed to Down or mountain sheep ; but the great weight, quickly and cheaply raised, together with the exceptionally heavy fleece, ensure a good return to the breeder and feeder. In fact, it is doubtful if any breed at a time when mutton and wool sell well realise so much on the year's feeding. Wethers of 12 st. dead-weight, cutting 16 lb. or more of wool, are not uncommon with ordinary management on good land. Rams have been known to yield a tod (28 lb.) of wool at a cutting, and ewes half a tod. The staple (or locks consisting of many fibres of wool arranged in natural bundles) of well-bred hoggs should be as broad as a man's two fingers, and may be up to 20 in. in length, with a bright lustre and wavy appearance.

The Lincoln has undoubtedly been used very much on some of the long-wool breeds in their period of making or improvement, possibly more so than the common use of the breed name of Leicester would suggest ; but in some of the other long-woolled breeds there is more suggestion of the Lincoln than the Leicester, though in the very early days of sheep improvement the Leicester practically had the field to itself. In the great sheep countries, south of the Equator especially, the Lincoln breed has played a great part, and there is still a big demand for them. The longest prices in British sheep—especially so over considerable numbers—are associated with the Lincoln breed. The sale of Messrs. R. and W. Wright's Nocton Heath Flock, in 1906, numbering 950 animals, for, it is said, over £30,000, stands out prominently in British flock records. The flock had a record of careful breeding since 1790.

The demand for mutton of finer quality, with less fat, has for a considerable number of years induced many farmers to use Down rams on the ewes, and where, in the memory of many, a Down-faced sheep was unknown, very large numbers of cross-bred sheep are met with. But on its native land it seems essential to keep to the Lincoln stock as a basis for crossing, and both for home and foreign use, heavy stocks of pure breeds are required.

The Flock Book was established in 1892. Care is taken to maintain the constitutional vigour that the breed inherited from the old Lincoln ; yet, in spite of the great weights attained, the plentiful bone, and the full woolling, there is a very noticeable aspect of kindliness about the breed.

THE WENSLEYDALE.

The Wensleydale is named after the Yorkshire Dale of that name, and is a modern appellation. It has much in common with the long-wool breeds of the lowlands, but has a distinctive blue colour in the skin of the face, ears and legs ; and often to a considerable extent over the body. This colouration is approved because the breed is largely used to cross with Scotch Black-faced sheep, producing what in England are commonly called Mashams, and in Scotland, Yorkshire crosses. The breed has considerably improved in one's own recollection, though it is not always attractive at first sight, as it does not evince so much as some other breeds the characteristics of early maturity. At the same time, it is a well-fleshed animal. It possesses much of the Leicester blood, and apparently had origin in the old Teeswater breed. It is recorded that the present type was established from a celebrated ram, Blue Cap, born in 1839, which had a dark-blue head, and a nearly black skin—and this view is commonly accepted. The better quality of its meat over the mutton of most long-wool sheep is, however, suggestive of a dash of hill or heath strain, though this probably comes from the practice of turning the sheep on to the hills, in some places, where they get a distinctive herbage.

The Wensleydale is a very big sheep, long and rather rakish as compared with other long-wools. Its long, strong neck, and pertly set head, suggest this. It is largely employed for crossing purposes in the north, and in parts of Scotland, particularly with the Black-face. It matures somewhat slowly, and many of these cross-breeds, known as Mashams, are fed down the Eastern Counties, and not a few reach the London area, where they are considerably run in the parks and on golf courses. Some get further south than London. The meat sells well, and a good fleece is cut. Some little confusion arises through the alternative name of Yorkshire crosses as applied to the cross in Scotland, the term " Mashams " not being commonly used in that country. The pure breed and the crosses are hardy and active, and I have found the crosses do very satisfactorily on the Sussex Weald. Some care, however, is needed to get them out at the right time of the year. The Wensleydale breeders are divided slightly in their views, with the result that the breed is represented by two Flock Books, published respectively by the Wensleydale Long-wool Sheep Breeders' Association and the Wensleydale Bluefaced Sheep Breeders' Association—a somewhat unnecessary attention to pay to the breed—where all might have been incorporated in one, with the advantage of strength that comes from union.

The characteristic features outlined by the Wensleydale Sheep Breeders' Association are : Face, dark ; ears, dark and well set on. Head, broad, flat between ears. Muzzle, strong in rams. A tuft of wool on forehead. Eyes, bright and full. Head, gaily carried. Neck, moderate length, and strong. Wool, bright lustre, curled all over body, all alike in staple. Legs, straight, and a little fine wool below the hock. Forelegs, well set apart.

THE COTSWOLD.

The Cotswold breed is native to the Cotswold Hills, which run through the eastern side of Gloucestershire in a south-westerly to a north-easterly direction. It is a long-woolled, hornless breed, with face generally white, though a slight greyness is not objectionable, as it is an old characteristic of the breed, which included white, grey, and mottled faces. As an indication of hardiness and indigenous character, a slight greyness is not to be despised. The breed can be traced back for several centuries, though at the end of the eighteenth century and a little later there was a considerable introduction of the Leicester blood, which worked well, and brought it into more accord with modern breeds, without impairing its hardiness. It is a big sheep, and it is exceptional that a long-woolled race should naturally have been established on such high, bleak hills as the Cotswold; though the fact that it was customary to house the sheep in cots or cotes at night in winter, made a considerable difference to the amount of hardship they had to undergo. The practice of providing these cots on the exposed hills or wolds was so general that the hills were named after them.

In its unimproved condition the sheep was large of frame, coarse, and a slow feeder, to a large extent finding its living on the short, sweet herbage of the hills. It is still a big sheep, one of the very biggest, cutting a heavy fleece of long, open, curly, somewhat coarse wool. Hoggets produce up to half a tod (14 lb.), but a general flock average is about 9 lb. Much of the Cotswold district gradually came under the plough, and the sheep, in its modernised state, was brought more under the influence of folding, both on seeds in summer and roots in winter. The modern Cotswold is long, with a notably straight, broad back, carried well out to the rump, with the ribs well sprung. Improvement has been made in the under line, which, with long legs and a somewhat cut up flank, used to spoil the outline, and did not recommend the animal to those who were accustomed to something that filled the eye better. The wool has also improved in quality. The mutton, in common with most other long-wool breeds, is not of choicest quality, being coarse in fibre and lacking in flavour as compared with the short-wool breeds. There is, however, quite a distinctiveness about the breed in its carriage and contour. The head is well set on, the face is characteristic, with a full tuft of wool which sets out boldly and long.

Without complimenting the breed at every point as being suitable to go on to a great variety of soils as a pure race, there is no doubt as to the great value the Cotswold possesses for crossing, and of the great part it has taken in the formation of some of our most valued mutton-making breeds. It has crossed so well with the

Downs, that crosses with them have largely occupied the land once held by them—being encouraged by the increase in arable farming. The big Down breeds of the south and west of the country owe many of their good features to the Cotswold, the Shropshire, Hampshire, and Oxford especially benefiting. It may be taken as a very fair axiom that for sheep of Down character to winter well on roots in the close fold on cold, wet soils, some Cotswold blood must be present in a greater or less degree.

The Cotswold played a great part in the formation of the Oxford Down, as it had, at a more remote period, and in a much less degree, in the Hampshire. The Oxford is the product mainly of the crossing of the Cotswold and Hampshire, and experience shows that the Oxford is the typical sheep for wintering on arable land on the colder oölites; moreover, the success of Oxford crossings with other breeds as compared with other crosses, is noticeable largely on those soils, and other similarly cold ones. The Cotswold, in its old indigenous condition, built up a constitution that enables it to winter under cold, wet conditions as none but hill sheep can; and no matter how small may have been the amount of Cotswold incorporated into a type, that sheep will winter better on cold lair than without the Cotswold influence. I had under my almost daily observation for fully a quarter of a century, flock experiments conducted for practical purposes by one of the most gifted sheep breeders I have known. His soil was light, medium, to heavy on the oölite in Bedfordshire. That county has no indigenous breed, its old heath breed having long been crossed out, though I can just remember throw-backs to it in an occasional speckled-faced sheep in carelessly bred flocks. He rang many changes in the course of years, but as he worked away from the Cotswold cross too far, he found that the sheep suffered in hardiness. He could work through to almost a pure Hampshire, which was his aim, but on each occasion, as he got nearly to purity, he had to bring in a fresh dash of Oxford to get the Cotswold influence. This influence had been marked even in his earlier experiments on Leicesters as a foundation stock. Yet within twenty miles on the Hertfordshire chalks Hampshires wintered perfectly; in fact, would make another stone to a stone-and-a-half more on a winter's feeding over drafts from the same lots sent on to the Bedford oölites. It is not impossible that other crosses would not give some of the necessary features to meet the cold, wet folds; but in working through the Downs I have not witnessed it. The appreciation of the Cotswold for crossing with the Suffolk when away from its driest lair, is shown by the fact that some good old established flocks exist in the Eastern Counties.

The aptitude of the Cotswold to thrive in wet, dirty folds is often attributed to the clodding or bouldering of dirt that gathers on the underline and down the legs, as it is thought that the earthy cloggings form a pavement which carries the carcase above the mud; to some extent this does occur, but it is least effective when the ground is in a miry condition, and with sheep which have strains of Cotswold in them, but the wool of which has not been appreciably influenced, and there is no special clogging, It is probably rather to constitutional influences than to mechanical ones that they do so well in wet, dirty pens. The Cotswold is a small breed numerically, but it has a value for crossing and reinvigorating breeds, which it would be unfortunate to lose. The Cotswold Flock Book was first published in 1891.

The sheep fatten out at from ten to fourteen months at 20 lb. to 25 lb. per quarter, though very great weights can be obtained by liberal feeding. A good general average of wool for tegs to cut is about 9 lb., and the wool should be long, lustrous, and thickly set, of medium fineness, with a well-defined lock, showing a wavy curl; in tegs this should measure from 10½ in. to 11 in.

THE KENTISH OR ROMNEY MARSH.

This white-faced, long-woolled breed is derived from the indigenous breed of the Romney Marsh, and is found in a highly improved state. It combines, in an exceptional degree, good mutton and wool characteristics. As the breed is dealt with somewhat fully in the section on the "Management of a Grass Land Breed," the references here are given briefly.

The following description of the typical Kent or Romney Marsh sheep is as follows: Head wide, level between ears, with good thick fore-top, no horns nor dark hair on the poll, which should be well covered with wool. Eyes should be large, bright, and prominent. Face in ewes full, and in rams broad and masculine in appearance. Nose in all cases should be black. Neck should be well set in at shoulders, and strong and thick, and not too long. Shoulders wide, well put in, and level with the back. Chest wide and deep. Back straight, with wide and flat loin. Rump wide, long, and well turned. Tail set in almost level with the chine. Thighs well let down and developed. The fleece should be of even texture, and of a good decided staple, from fore-top on the head to end of tail, and free from kemp. The face should be white, and the skin of a clean pink colour.

DEVON LONGWOOL.

The Devon Longwool originated from the Southern Notts, or Bampton Notts (or hornless), a long-woolled breed of Devon and Somerset, white-faced, and hornless. It was very much crossed with the Leicester and Lincoln in early days. Another Devon long-woolled breed was found west of Honiton, and this was known as the Southam Notts (or hornless), which affected South Devon and Cornwall, and which is now known as the South Devon breed. This, then, had brown face and legs and a long fleece. Some confusion sometimes occurs between these breeds, as they are both entitled to be called longwools. In fact, indiscriminately, both have been called Devon Longwools, though locally there has been a distinction between the Bampton and the Southam breeds or strains. However, more definite distinction was set up when the Devon Longwool breeders established a Flock Book in 1900, and the South Devon men published theirs in 1904. The difficulty in respect to the naming and identifying of these breeds, after improvement was started, was due to the fact that there was not a constant name ; and the long-wool breeds of Devon, owing to the very free use of Leicesters, were much associated with that breed. At first they were known as Devon-Nott, but as the Leicester type increased, the name Leicester Longwools began to be applied some sixty years ago, because, owing to the large infusion of Leicester, they were disposed to regard them as a sub-variety of that breed. But, a few years after, when the breeders dropped fresh introductions of the Leicester, and more definitely selected the sheep within themselves, establishing a marked type, they took up the name of Devon Longwools. The Bamptom and the Southam were distinct in type, and breeders in the respective districts, recognising what features proved most successful on their land, have since bred to those points, and without the introduction of new blood have evolved the distinctive breeds of Devon Longwool and South Devon. This distinction is doubtless seen by many breeders throughout the century or more of evolution, but not appreciated to a similar extent by those outside.

The Devon Longwools are noted for their size, though this has been judiciously modified by the breeder. Still they rank among the heaviest breeds in this country. They show the good features of the long-woolled breeds generally, and cut a fleece of long wool of good quality. The breed does remarkably well on the rich pastures of Somerset and Devon, is strong and hardy, produces a good crop of lambs which are easily reared.

SOUTH DEVON.

The South Devon originated from the Southam, in the district affecting South Devon and Cornwall. Some features of its origin are mentioned in the account of the Devon Longwool breed. The South Devon sprang from one of the strains of rich lowland or marsh pasturage, which included the Marsh or Fen breeds of Lincoln and Romney Marsh ; in themselves coarse and big, but under suitable crossing at a distant date, and by good management and selection, now long-wool breeds of first importance. They were of coarser type than the long-wools of the pastures at higher and drier levels. The South Devon originally had somewhat brown faces and legs, and the long fleece of the Marsh breed. The type has been judiciously handled, and the formation of the Flock Book Association in 1904 has given considerable impetus to its interests. The sheep run to big weights, and possess the faculty of early maturity in a marked degree. Lambs shown at Smithfield in 1909 gave an average daily increase from birth of 13 oz. to 51 oz., which is the highest recorded at that Show. With a good constitution, it is vigorous and thrifty, and does well in the fold or at grass, being able to hold itself together on hard fare and in exposed places. The fleece is curly and dense, of long, staple, lustrous, and clean.

Scale of Points for Standard Type of South Devon Sheep.

General character and appearance	14	points.
Head, wide, well covered with wool 6		
Face, white, full, rather long, muzzle broad, nostril dark and open 3	12	,,
Eyes, large, bright and prominent. Ears fairly long, medium thickness, covered with hair, preferably with black spots on white ear .. 3		
Neck, strong and big at base, well set on at the shoulders, medium length	6	,,
Shoulders, well set, the top level with the back and base of neck	4	,,
Chest, wide and deep	3	,,
Back, level, with a wide and well developed loin	10	,,
Legs, full, well let down, with a deep, wide twist	8	,,
Body, deep and long, ribs well sprung, thick through the heart, with fore and hind flank well developed	7	,,
Rump, wide and long, and even with the loin	4	,,
Tail, large, set on almost level with the loin	4	,,
Wool, dense and even, great length of staple, covering the whole body, curly, and free from kemp or hair ..	14	,,
Skin, mellow and of pink colour	4	,,
Carriage, smart, legs straight, of medium length, well let on outside the body, bone large	10	,,

100 points.

The Roscommon.

The Roscommon is the long-woolled breed of Ireland, and originated from a coarse native race such as commonly inhabited the rich lowland and marshy tracts. Ireland, as a country, is excellently suited for sheep farming, and has always suffered in its sheep industry, as it is mainly under small holdings. Experience shows that sheep farming never thrives under a system of small holdings. Ireland, especially the western province of Connaught, where sheep are more numerous than in the north, was much associated with the Roscommon breed, and in the west the winter grazings have taken very much the nature of commonage. Twenty years ago I found that often the walls were thrown down in the autumn, and the owners dare not replace them, for, as they put it, " the boys would make it hot for them at the next fair," as their commonage was hindered. The breeding was, and practically always had been, left to chance rams, which was altogether opposed to improvement in the sheep stock. Where bigger holdings prevailed, better management was the rule, and with good pasturage and less neighbourly interference of a prejudicial nature, a better sheep was found. But the breed was in a very backward condition until within comparatively recent years, and possessed many of the features that Culley, at the end of the eighteenth century, attributed to it in his description as follows :—

"I am sorry to say I never saw such ill-formed, ugly sheep as these (referring to 95,000 he saw pitched at Ballinasloe Fair) ; the worst breeds we have in Great Britain are by far superior."

He further said, they had nothing but their size to recommend them. They possessed very long, thick, crooked, grey legs ; heads long and ugly, with large flapping ears, grey faces, and eyes sunk, necks long, breasts narrow, high narrow herring backs, hind quarters drooping, and tail set low. A bad enough record indeed ! Some little improvement, the result of English breeds, was noticeable occasionally. Systematic improvement was, however, slow ; and this was not well established until about 1850, from which time, however, great advance has been made. To a few men in Co. Roscommon the chief improvement is due, though not entirely, but owing to the more general improvement made in that county, the breed has been established as the Roscommon.

The Leicester has been mainly used to improve the type, and as on all other long-wools it very quickly wrought improvement, the Roscommon felt its influence. Having been in the hands of men with considerable skill in breeding who recognised the value of modern principles, it has come much more into line with other long-wool breeds. The Lincoln, however, has been successfully used, and has more modernised it. The meat and the carcase are improved, and it is no longer necessary to keep them three or four years before fattening them. They come out at big weights, but the wool is an especially good feature, being soft, rich and full, good hoggets giving 10 lb., and average flocks 8 lb. Very great weights have been grown, some rams in high-class flocks up to as much as 24 lb., and ewes from 14 lb. to 16 lb. To the Flanagans, Flynns, Cotton, Roberts, Taafe, and Blood Smyth (Co. Limerick), the breed owes much of its early improvement, and, in some instances, its later.

The Roscommon is an excellent grazier. Moreover, it crosses well with the Down and Black-faced sheep. Visitors to Ireland who knew it twenty-five years ago cannot fail to notice how much cross-breeding there has been, and it is very striking how this prevails even in districts and in climates where it would probably not be expected. Good early maturing lambs are raised from the cross, the strong mothers supplying plenty of milk. Like the other long-woolled breeds of these islands, it is a hornless breed. The face is long and white, the indigenous grey of the faces, as well as of the legs, having been bred out.

There is usually a tuft of wool on the forehead ; the ears have been modified to a finer texture and moderate length. The sheep has an excellent constitution, and is finding some support abroad. Moreover, it could be brought quite in line with other long-woolled breeds without unduly sacrificing its vigour. The endeavour on the part of some of the early breeders to depend too much on selection, and avoidance of outside help, retarded its development, and made them slower in bringing forward the breed. With an instillation of more advanced blood, a foundation was formed on which to work, and breeders can now keep clear from new introductions. The old idea that it is necessary to retain only indigenous blood in a breed is exploded. It is advantageous to keep it in in a marked degree where the sheep are required to remain under very similar conditions to those under which the breed characteristics were formed ; but with a variation of these, and a desire to alter them, it has rarely paid to adhere to only one breed, although, after admixture, selection may do best without further addition.

CHAPTER VIII.

CROSSES AND CROSS-BREEDING.

The grass-land breeds from the hills and lowlands adjoining are making a very pronounced influence on the sheep of the southern half of the country, where they are kept both in the pure and cross condition, and they are finding a home, especially where there is no stock of the indigenous blood remaining. Where, however, there is much fattening on swedes on arable land in winter, they have not been adopted as much as in those districts where more reliance is placed on grass feeding. In counties such as Bucks, Beds, Northamptonshire, and part of Cambridge, there is no indigenous breed left. The tracts of country of closely similar soil in those counties were too small to produce noticeable races, and they were shoved aside by the more highly-bred animals from more typically sheep districts which had assumed more fixed type. Local breeds were cast aside, and how the food influenced the changes was well illustrated to me quite a number of years ago by the late Mr. Peacock, of Stanford, Beds, who at that time was an octogenarian, and had been recognised for many years as a skilled breeder. His memory carried him back to the pre-Norfolk rotation days in that county. His earliest recollection was of the native heath breed, a white and speckled-faced horned sheep (which he described as being lean, gaunt, and unthrifty, reaching maturity only when three or four years old, and then looking more like a woolled donkey than a well-fattened sheep). The Leicester sheep followed as clover growing and root culture became more generally adopted, and these were subsequently ousted by the Improved Leicester. The general adoption of the Four-Course System called for a sheep better adapted to folding, and for a short time the Cotswold appeared, to be followed by the Oxford Down. Then the West Country sheep, of the type found at Illsley Fair, in which there was some Hampshire blood, and with an increase in catch cropping for spring feed, the definite Oxford-Hampshire cross was more adopted. Not a few pure Hampshire Downs have been kept, though as breeding stock they have not proved a success on the colder land, yet on the chalks they have been quite successful.

But it must be borne in mind that whilst a breed may not be suitable to be permanently kept as a breeding flock, it may be quite satisfactory for fattening out in a few months' feeding, as in the latter case there is not time for much alteration in type, and, of course, no breeding is done. Then there is no need to think about the perpetuation of breed features.

With respect to the Oxford and Hampshire cross suiting the counties mentioned, it is doubtless due to the greater infusion of Cotswold blood purveyed through the Oxford Down ; for wherever the Down is kept on the oölite it seems to be imperative that there shall be some appreciable quantity of Cotswold blood. Sheep with Cotswold blood in them appear to be warmer on these cold soils in wintering on arable land. But there seems to be a decidedly greater adaptability of Western Downs over Downs from other districts, such as the Shropshire or Suffolk. Moreover, in grass feeding, the Lincoln and Leicester long-wools do not thrive in their pure state as well as they do farther north. The Oxford Down has considerable influence over these districts, and this is undoubtedly due to the fact that they produced, as a result of their natural features, indigenous breeds of the heath type. The results as a rule obtained from the breeds employed generally, are satisfactory ; but in the general resorting of breeds that is going on, such districts as these having no original extant breed would stand a good chance when adopting some of the hitherto less common types.

Two southern races which have come into great popularity in recent years—the Romney Marsh or Kentish, and the Exmoor— were considerably influenced by the Cheviot many years ago. The Exmoor was dealt with by intentional importation, the Kentish through the accident that, in the pre-railway days the Cheviot sheep were brought by water to London, and when they met a bad trade, and it was desired to avoid the river and market dues, the sheep were landed on the Kentish coast, and bought by farmers. Gradually, they mingled with the native sheep, and impressed on them some of the heath features ; greatly improving the flesh, and making it quite distinctive from that of the other long-wool breeds of the rich lowlands and fens, or marshes. The cross between the Southdown and the Romney Marsh sheep is very popular, and extensively made, with a decided tendency to further development. Other crosses are made with the bigger Down breeds, but they only occasionally find popularity, though as producing bigger lambs for early fattening, they are preferred by many. A common verdict is that the Southdown cross is better for fattening out as wethers, as so many in the South-eastern Counties are, but that the bigger Down crosses should be killed fat as lambs, as local graziers do not appreciate them, and butchers prefer the Southdown wether cross. Long-wool crosses with the Romney sheep are not generally popular with the Marsh men, but Mr. Hobbs, of Brookland, one of the biggest flockmasters—in a district where flock masters keep many thousands of these big sheep—tried the Wensleydale cross with considerable success. As lambs, he was much dissatisfied with them, but as wethers, they made a considerable sensation in the Rye Market two years ago, where they sold at very high prices, the butchers showing great keenness in buying after the first batch had been killed, and their properties were recognised.

Hampshire crosses are common through many parts of the country, and they have effected a great work on the Lincoln's during the past quarter of a century, since coarse meat has been difficult to sell, except at times of mutton scarcity. For the last year or so all mutton has sold well. The Hampshire, crossed with the coarser-fleshed long-wools, makes the mutton far more palatable, and although the fleece is lessened, the total value of the sheep is increased.

From the carcase aspect, the Suffolk and the Cheviot cross, and crosses of either of these with other breeds, are most successful, although, as a pure-bred, the Suffolk is individually the most successful race. The Suffolk, in which there is a considerable dash of Southdown, is the sheep of highest quality to-day. The successful Suffolk is the sheep of the eastern side of the county, near the sea, where the truest Suffolk Down is found. In west Suffolk, what may be described as a too-quickly-get-there system of breed improvement was in vogue some years ago, and the Hampshire cross was brought in for this purpose, and although rapid improvement and a valuable sheep resulted, it missed the distinctive character of the Suffolk as prized to-day. The modern leading flocks produce good carcases of excellent lean meat, and it is this kind of sheep that crosses so effectively with other breeds to improve the meat. The carcase competitions at Smithfield have done much good in calling attention to the quality of meat ; and especially of the relative proportion of lean to fat, and the Suffolk, and the Suffolk-cum-Cheviot are setting a good example.

The Cheviot was used on the Exmoor many years ago to improve it when it was comparatively an insignificant breed. Now the Exmoor is, in many places, ousting the native breed on the Wiltshire Downs ; whilst the Cheviot itself, as well as the Scotch Blackface, are also actively pushing the native sheep aside on the soils other than chalk. The Cheviot comes in again in the Midlands, where the Scotch-bred Leicester-Cheviot is pushing its way. As one parent in the great Border Leicester breed, the Cheviot shows its personality, whilst in Wales it has had great influence on the most improved of the Welsh breed, having impressed its character on what was previously a very light fore-quartered breed, and having much forwarded its maturing. The Cheviot has been more valuable in its crosses with the Welsh sheep than the Black-face. In fact, the Cheviot stands out as the breed which is influencing modern breeds of sheep far more than any other, though the Black-face is establishing itself, both as a pure breed and a cross breed, over a very extensive and varied country. But this is more a matter of direct substitution ; the large amount of crossing long-wools with one or other of the Down breeds affects mutton production the most at present. The Black-face has done remarkably well on the Irish breeds. Being by origin a heath breed, though altered by environment, and partaking many characteristics of more truly mountain breeds, it mates well with other types in which there was heath origin. But it also mates well with the longer wools when judiciously handled. The place of the Black-face in crossing is likely to be a greatly extended one. As it merges in other breeds, and is kept in climates, on soils, under different treatment and feeding, to those to which it has been subjected for centuries, there will be doubtless many changes in its characteristics. It is not always wise to fight against them when the environment is changed. To keep a pure breed under totally changed conditions to those under which it developed, and then expect it will maintain the features commonly associated with it, is ridiculous. It must change, and when the change means adapting itself to new conditions, it is doing what it ought to do ; and what are known as breed characteristics, in such circumstances should not be insisted upon. Whether it is of best economic value for breeders of Southdowns at a distance from Sussex, as on the Cambridgeshire chalks, to fight against the natural changes which the local environment insist shall occur, merely for the sake of keeping them to the Flock Book standard, is very debatable. The Black-face has a great work in being used on some of the breeds which have been developed on fat-making lines. It can be a splendid infusion in recasting breeds.

The Lincoln has proved itself a fine sheep for the county to which it has long been attached. It did great work on many long-wool breeds, and, as showing that it has possibilities that ought not be ignored, it was the one important breed that was used by the pioneers in the improvement of the native Cheviot. It converted a breed of many good points and several inferior ones into the prominent race it has become. It is not surprising, therefore, that it mates well with other heath or (as they were evolved by the aid of the Southdown), as they are called, Down types, to produce a big carcase of meat and a good fleece of wool, suitable for the rich pasturage of the land with which they are identified. It is probably true that there is practically no race that did not feel some effect from the Leicester, and it even touched the Southdown in the earlier days, and doubtless among other features improved the fleece, which previously was very short and light. At times some of the Long-wools require a fillip from the Leicester to maintain their characteristic features and constitution; and although the Leicester itself may have received a stiffening in return, it has been well directed, and its place among the wool-producing breeds is necessary. Many parts of the North of England, and much of the Scottish Borderland, and still further inland, would have fared badly but for the Border-Leicester, the Leicester and Cheviot breed (for although the Border-Leicester men are advocates of the narrow or one race breed, and urge there is nothing but Leicesters, their case is weak); and cross breds with the Border-Leicester are to be met with in almost all parts of the country, especially as a great number come to be fattened as far down as the Thames, and occasionally south of it.

It is impossible, within reasonable space, to specify where all cross-bred sheep are located. Some are kept constantly in some districts, and in others only occasionally; but the number of crosses kept is very great. It is a great feature in the sheep management of these islands that careless mongrel breeding is steadily diminishing as breeders better realise the advantage of crossing with pure types, that is, with those bred straight, skilfully and carefully for many generations. The pure bred is the foundation of the cross bred as opposed to the mongrel; and the necessity for maintaining as many pure breeds as will embrace the whole of the good indigenous characteristics which were evolved during centuries must be obvious to all. However, with changing markets and conditions, it becomes necessary, just as it has been during the century or century and a-half that the breeds that are with us were modified as circumstances directed, should be recast or remodelled, to meet conditions as they alter. That this will be done there can be little doubt; and that it will be done with all skill is equally certain, even though it may call for bold men to grapple with the problems that arise in their individual localities.

Those interested in sheep have a very interesting period in front of them in regarding the changes as they will be made. Some districts have never yet been satisfactorily provided with a breed, and new ones will be found for them; or older ones will be infused with other blood, and more suitable ones will be evolved from them. In a sense there is no best breed. Most are the best breeds for the districts in which they have been produced, but there is no universally best breed. Advocates for the type to which they are attached often conscientiously believe, from their own experience, that none equals theirs, and therefore should be universally adopted. But that is rather a parochial way of regarding the great variety of conditions this little island provides, and the extraordinary number of distinct features, characteristics, and uses contained in the sheep themselves.

The following useful information was published by the Board of Agriculture and Fisheries, and from its specific nature, it is suitably given here :

CROSS-BREEDING FOR MUTTON IN THE NORTH OF ENGLAND.

South of England buyers often visit the great autumn sheep auctions of the North, in order to purchase cross-bred lambs and draft cross-bred ewes. Some information as to the manner in which these cross-bred sheep are produced may, therefore, be of value to farmers and others who are interested in the production of first-class mutton.

The greater part of the mutton produced in the four northern counties of England is cross-bred, and the same may be said of that fed in the South of Scotland. The chief reasons for the method of breeding prevailing in these districts are to be found in the mountainous character of the country, the long and often severe winters, and the special suitability of the climate for turnip-growing.

Mountain Breeds.—All the crosses met with in the North of England have their foundation in the mountain breeds : the Cheviot, on the low and verdant Border hills ; the Black-faced mountain sheep (Scotch Black-face), on the higher hills of Scotland, and on the Pennine chain and its spurs running into Northumberland, Cumberland, Durham, and Westmorland ; the Herdwick, on the poor mountain land of Cumberland and Westmorland ; and the Limestone fell sheep of Westmorland.

Of these the Herdwick is the hardiest—possibly the hardiest sheep in existence—and able to get its living throughout the winter on the scanty herbage of the fells, so long as the ground is not covered with frozen snow. Closely following the Herdwick for hardiness are the Black-faced mountain and the Limestone sheep. Without these sheep very little fell farming would be possible, and there would be no means of profitably turning to account the mountain herbage in these districts. But they are small sheep, coarse in the wool, slow in maturing, and too wandering in habit to settle down quietly to feed in small fields and folds ; consequently, as distinct breeds, they are not profitable for stocking tillage farms, which are comparatively highly rented, and on which the production of rapidly maturing lamb and mutton is aimed at and quick returns are expected. On the other hand, all three breeds are renowned for the large proportion of lean meat in the carcase, and for the sweet and fine-grained quality of their flesh.

The Cheviot has a fleece of fine quality, and is a much tamer sheep than those just described. It is very compactly made and yields mutton of the highest quality. Of the pure mountain breeds it is certainly the best adapted for fattening on the lowland farms, though it is small in size.

First Crosses.—When these mountain sheep are crossed with any of the large-sized quick-growing breeds, they produce lambs of excellent quality, quickly maturing, and very profitable, either for the butcher or for breeding from as cross-bred ewes.

Border-Leicester—Cheviot cross.—For the first cross the Border-Leicester ram is the one most in favour for use on these mountain ewes in Scotland and the North of England. Like all sheep of the long-woolled breeds—Border-Leicester, Leicester, Lincoln, and Cotswold—it carries far too great a proportion of fat in its carcass ; but it is a large, early-maturing sheep, with excellent fleece, and begets good-backed lambs that both grow and fatten rapidly. It has also the important recommendation of having a narrow head, which is inherited by the lambs, and so the difficulty to the small mountain ewes of lambing large lambs is not materially increased by its use as a sire. Where rams of the Down breeds are used, lambing difficulties and losses may occur, owing to the large heads of the lambs.

South country farmers frequently raise the objection to the Border-Leicester—Cheviot cross that the flesh is " sappy " and does not keep well, but this is probably through associating this half-bred white-faced sheep with the pure Leicester and other white-faced breeds that carry so much fat. As a matter of fact the Cheviot so satisfactorily corrects the inferior quality of the flesh of the Border-Leicester as to make the cross one of the best mutton carcasses.

The ewes of this cross are handsome, compact sheep, of good size, with fleeces of the best quality ; they inherit the good milking qualities of the Cheviot, are free yeaners—generally bringing twin lambs if in good condition at tupping time—and easily fatten while suckling a pair of lambs. Very large numbers of half-bred lambs and draft ewes of this cross are sold annually at the autumn store sheep sales of Scotland, a large proportion of them coming across the Border, the lambs for winter fattening and the ewes for early lamb-breeding.

Border - Leicester — Black - face cross.—The Border - Leicester —Black-face cross prevails in the adjacent districts of Cumberland and Westmorland, of which the market town of Penrith is the centre. Some thousands of these " Grey-faced " lambs and draft ewes, as they are locally named, are sold in the Penrith Auction Mart every autumn. These sheep are not so compact in make as the half-bred white-face just described, nor do they carry wool of such good quality ; but for good feeding qualities and high-class mutton, with plenty of lean, they would be difficult to beat. The ewes have the excellent milking qualities of the

Black-face, and mostly drop couples. Early lambs of this cross, having a little colour in their faces, take the market well.

Border-Leicester—Herdwick cross.—For high-lying tillage farms, the Border-Leicester—Herdwick cross sheep are excellent ; their mutton is of the very best, they are hardy and good sized and thrive well, but the ewes are not so prolific as those of the other two crosses, while the fleece is coarser.

Wensleydale—Black-face cross.—A very favourite first cross along the adjacent Westmorland and Yorkshire borders is that of the Wensleydale—Black-face. Sheep of this cross also go by the name of " Grey-faced." The lambs are rapid growers, and the mutton is of high repute, but they fatten more easily when nearly full-grown than as young lambs ; they are therefore better adapted for the mutton market than the lamb market. The Border-Leicester sire certainly scores over the Wensleydale as a producer of fat lambs ; but the latter has a special value as a sire for a second cross.

Second Crosses.—Coming now to the second cross, the Border-Leicester ram has not much advantage over the rams of other large breeds in the matter of begetting lambs that come easier to the birth ; for the first-cross ewes above described are large and roomy enough to give birth quite naturally to fairly large-headed and wide-shouldered lambs. Rams of the following breeds are used on these half-bred ewes : Border-Leicester, Oxford Down, Wensleydale, Shropshire Down, Leicester, Lincoln, and Suffolk Down. The first-named was at first much more largely used than all the others put together ; but the Oxford Down, especially for white-faced cross ewes, is coming rapidly into favour, as may be gathered from the large and increasing numbers sent in recent years to the great ram sales of the North, at Kelso and elsewhere. The plump, dark-faced, close-coated lambs of this Oxford—Border-Leicester—Cheviot cross are great favourites with the butchers, and carry more lean than lambs produced by the use of the Border-Leicester ram a second time ; and even on the grey-faced ewes, except when fat lambs are required, the Wensleydale ram is to be preferred to the Border-Leicester. In fact, the heaviest cross-bred sheep produced in Cumberland and Westmorland are those of the Wensleydale—Border-Leicester—Black-faced breed. For several years in succession at the Penrith Christmas Fat Stock Show the 1st prize pen of shearling wethers were thus bred. These sheep have averaged 230 lb. live-weight, and have realised 84s. each.

A common practice on the higher arable lanbs of the Border district is to cross the half-bred ewe with a half-bred ram, bred on identical lines. Fattened during the winter, the lambs of this cross produce carcasses of solid lean mutton in much favour with the butcher.

CHAPTER IX.

SELECTION OF BREED.

The selection of a breed suitable for a particular locality is dependent on many conditions. The first and most important is the soil, after which come the climate, system of farming, nature of herbage, and purpose for which they are required. Down sheep are not best suited for rich pastures, and long-wool sheep are not profitable on Down land. To a less degree all other sheep are influenced by the soil on which they are carried, the extent being regulated by the dissimilarity of the soil to that to which they are indigenous. A very successful exhibitor of Shropshire sheep in Nottinghamshire divided his flock, one portion being kept at home on a good loam, while the other was sent but a few miles away to a sandy soil; the result was that he continued to win prizes on the loam, but on the sand the flock lost size and characteristic features, and so deteriorated, notwithstanding every effort to prevent it, that in a few years he was obliged to substitute another breed for them. Hampshire Downs taken on to gravels and cold loams quickly lose their type, and are not so profitable as on their native chalks, although the sheep may be moved but a few miles. Much of this is due to the colder lair in winter, as in summer time they do well. With other breeds the same variation in type is experienced when they are removed from conditions under which the breeds are built up. It is therefore important to regard the nature of the soil when entertaining the idea of making a change.

There is, however, one important point to be remembered in the selection of a suitable breed for a particular farm. Although the conditions may not be favourable for breeding, they may be for feeding. When breeding, it is usual to keep to a selection of the breed for a number of years, during which the sheep gradually change their type; whereas during the few months when imported sheep are being fed little change takes place, and if it does, it need not be of great moment. As a matter of fact, the greater portion of the Down-bred sheep are fattened off on land at a considerable distance from their breeding place, and thrive well.

Soil and Climate.—Climate, undoubtedly, has an effect on the thriving capabilities of sheep. Those accustomed to dry lair, a moderate rainfall and mild climate, suffer when the conditions are changed. It is noticeable that the indigenous breeds of the southern portion of the country, where the rainfall is light, the climate mild, and the lair is dry, are almost, without exception, short-wool breeds. An exception is found in the Romney Marsh sheep, but exceptional conditions account for this: they are indigenous to the rich low-lying marsh land of a small tract in the comparatively dry climate of Kent. The proximity of the sea, and the low-lying position of the land, however, counteract other influences, and a long-wool breed is found. Dryness of the climate has an effect on the herbage beyond that which is caused by the nature of the soil. Down land naturally carries short herbage, but the herbage of the Downs differs from that of the thin soils of the limestone situated in more northern and wetter parts, where a different type of sheep is carried.

A great effect of moisture on land is that sheep which are indigenous to it, or have been carried on it for a long period, open their claws or digits when they tread on it. This is necessary to prevent them from sinking deeply, whereas on dry soils the feet open very little. Thus, on the Downs, where the ground is rarely sufficiently moistened to allow the sheep to tread through the turf, the feet are small and the digits close, as compared with those of sheep carried on loose soils supporting rich pastures.

However, the Suffolk breed, bred in the driest climate of Great Britain, has a comparatively open foot, in spite of the dry land on which it runs, but it originated in coarse heath, and the claws took a spreading habit—accounting for their comparative freedom from footrot.

The formation of the feet has an important bearing on their liability to foot-rot. Sheep which for generations have had to expand their feet have developed a hard skin between the digits, and this is not easily abraded when it becomes moistened, as when brought into contact with long wet herbage, or when the sheep are penned on wet, gritty, arable land ; whereas under similar circumstances the close-toed sheep are very susceptible to fracture of the skin. When once the skin is broken, the germs of the foot-rot disease easily establish themselves, and the foot becomes diseased. A difficulty is, therefore, experienced in placing Down sheep on soils carrying rich pasturage, or on arable soils which produce friction about the feet. The suitability of the sheep for the purpose which is likely to prove most remunerative is an important matter. Generally, in these days of foreign competition, meat is the first object, as wool is imported at such a low price ; still, the value of a fleece which weighs 10 lb. or more is not to be ignored. The type of sheep must be decided to no small extent by the characteristics of the farm and the kind of cropping it will carry with most success. It is fortunate that while mutton from white-faced sheep is not so valuable per pound, the sheep produce large carcasses and abundant fleeces of wool, which together are of great value ; so the sheep-grazier is not altogether outplaced by the arable land farmer, who, owing to the greater expenses to which he is put in obtaining crops, requires some advantages to make matters equal.

CHAPTER X.

The Points and Nomenclature of Sheep.

Points of Sheep.—Before dealing with the management of sheep, it is advisable to discuss some of the points which should be looked for in them, as badly-bred sheep are rarely profitable, or, at any rate, are not so profitable as those better bred. In the first place, a sheep should possess features and an outline which are pleasing to the eye. The animal should have a well-balanced appearance, otherwise there is some feature lacking or too prominent. A lean sheep can never show to such great advantage as one which is fat, as meat fills up the frame, and makes it more even and level. It is not, however, necessary that an animal should be fat for its good points and thriving properties to be noticeable. There are signs of good breeding, rapid maturing, good wool, and growthiness in a well-bred sheep, no matter how poor it may be or how ragged the wool may appear, and these are readily seen by a good judge of animals. Features which are good in one breed are generally good in another, though, of course, there are characteristic features in every breed distinguishing them from other breeds ; but in broad principles that which is good in one must be looked for in another.

The *body* should have a well-squared appearance ; neither end should taper ; the hind quarters and the fore quarters should finish boldly, and the line of the back and that of the belly should be parallel.

The *tail* should be well set on in a line with the back, so as to give the appearance of finish. It should not be too high, and if too low it denotes slackness of the hind quarters and coarse breeding. The tail, or dock, should be broad, affording a good grip when taken in the hand.

The *loin* should be broad and flat. The sheep should " back " well. If the spine rises high about the hindquarters the sheep is unthrifty, as it is in an " unimproved " condition. It will generally be seen that where the backbone is raised high on the loin and rump, the girth through the heart is small and the forequarters are light. These are conditions which almost invariably are present in wild sheep and those which have not been subjected to improvement by careful selection and breeding. Almost all old illustrations of sheep show these characteristics. Broad, deep *forequarters*, and broad, flat loins indicate powers of early maturity, consequently they must be looked for before any other features. To those accustomed to flat-backed, white-faced sheep which carry a large amount of fat on the back, Down sheep appear to be narrow and high in the loin, as they do not lay on so much fat in that part unless specially fed. In comparing the touch of a long-wool sheep with a Down, a long-wool on which the backbone appears to be at all prominent is not ripe, whereas if the same amount is felt on a Down sheep it may still be in good condition for the butcher. Those accustomed to handling Down sheep must therefore be careful when handling long-wools, or the condition of the latter may be over-estimated.

The *neck*, or *scrag*, should be broad, as a thin neck is usually associated with a narrow forequarter. The neck should taper fully from the body and shoulder, and be brought up from a square, deep brisket. Where this is the case the sheep " meets one " well. The *brisket* should show distinctly in front of the fore legs when viewed from the side. The neck and head should make a bold, level sweep from the nose to the shoulder, giving the appearance of a well-curved and full crest.

The *head* varies considerably in different breeds, in regard to both shape and colour. In the Down breeds a fairly broad forehead and a broad muzzle are usually preferred, and, except in the case of the Suffolks, the wool should come well over the forehead and about the upper part of the jaws. A sheep " well-woolled " about the head, particularly on the poll, is less liable to injury from flies, which cause great trouble at times, especially if through butting or other causes the skin is broken. The *ears* should not be too thin or papery ; on the other hand, they should not be too thick, as this indicates coarseness of skin. The shape of the ears differs with the breed. There should be no folds of skin under the jaw, as " bottle-throated " sheep are coarse in the skin.

The *teeth* are an important feature, for on the power to graze or gnaw well depends a good deal the amount of food a sheep will get. Short, closely-set front teeth last longer than long widely-set ones ; they are less likely to break, and when old they do not let the grass slip between them so readily.

No *horns*, or rudimentary horns, are permitted on Down sheep, as they indicate a tendency for the sheep to revert to the unimproved type. Sprigs, or snags, as the rudimentary horns are called, are regarded as serious blemishes. The wool about the head in front of the setting of the ears may contain black hairs, but behind that the wool should be absolutely free from them, as black hairs away from the poll indicate want of selection, and a tendency for the sheep to revert. Coming back to the body, the *shoulders* should be full but slightly obliquely set ; at the top they should be level, so as to give the whole back line a straight and square appearance. If not well buried they suggest the unimproved sheep, and the filling-in behind the shoulders is not complete ; this makes a narrow girth round the heart—that is, behind the shoulders. Deepness through the heart is so important that it must never be disregarded. Both the *fore* and *hind legs* should be set on squarely, neither turned inwards nor outwards too much. A good leg of mutton is an important part of a sheep, as it is one of the most valuable joints on the animal. The legs should be full from all points of view, and the meat should come down well towards the hocks. Squareness of frame when viewed from behind is greatly increased when the legs join well below the tail. If they fork too high up, the legs are not well developed.

The *wool* should grow thickly on the skin, and should be fine in texture and free from dark hairs, except in a few hill breeds. Many breeds of Down sheep have been developed from heath breeds which, in an unimproved condition, have a large quantity of black, harsh hairs intermixed with the wool, especially about the hindquarters and the head. These should be bred out, because they are not only injurious to the wool, but indicate insufficient care in breeding and want of thriving properties. When much dark hair is intermixed with the wool on the thighs the sheep are called "breechy." However, hair on the thighs assists water to drain more freely from wool, and in wet, hilly districts this may be an advantage ; and in some breeds a little breechiness is not objected to as indicating the stronger constitution, which results from the absence of chilling which wetter wool promotes.

The *skin* should be clear and healthy ; in almost all breeds a clear pinkish skin is best. A dark skin is likely to produce dark wool. Sheep with dark skins are liable to produce snags.

The *underside* of a sheep should be well covered with wool. This adapts them for cold lair. Those with little wool below thrive badly on cold soils, particularly in wet winters.

Handling.—"Handling" sheep conveys the idea of touching them to see in what condition they are. The most important points to handle when dealing with ordinary farm sheep with a view to sale or purchase, are the loin, dock, neck, and scrag. The hand should be stretched across the loin, which will show its width and firmness ; the fingers should be drawn up to the spine to prove how much it protrudes. The dock should be gripped to ascertain its breadth and fatness ; and the "nick," a depression felt for a short space along the spine above the tail, should be found by the fingers, because its size denotes the fatness of the animal, as it is formed by the protrusion of fat on the sides of the spine. If the loin is firm and flat, the dock broad, and the nick well defined, the sheep will "die" well, as these denote that the animal is in a good condition internally, and that the kidney fat is well developed. The animal may be felt along the spine generally, but except in very highly-fed or show animals, it is not often that the back is flat right along to the shoulders, so that it need not be looked for in moderately-fed animals, although a tendency in that direction is advantageous ; at any rate, the greater the breadth there the better, as indicating both the animal's thriving properties and its condition. A grip between the thumb and the hand will indicate the strength and condition of the scrag. It is advisable to turn a few sheep to see what condition they are in on the underside ; a brisket well covered with meat indicates ripeness in the forequarters ; and if the scrotum of castrated sheep is well filled with fat it will prove good internal condition.

Dentition.—A great number of terms are employed to distinguish sheep at various ages. They are so many, in fact, that few farmers are accustomed to all, especially as those in common use in some districts are rarely used in others ; therefore, the enumeration of a few is advisable. As some of them are founded on the condition of the sheep's teeth at various times, the dentition should be understood. For the purposes of the farmer, the *front teeth* or *incisors* are generally sufficient, as it is rare there is occasion to look to the molars or back teeth. But to the exhibitor it is sometimes an important matter to refer to them. There are no incisors on the upper jaw, but in the place of these is a hard elastic pad, as in the case of the ox. The farmer is content to know that the temporary incisors remain in their places until the sheep is rather more than a year old, when the central pair are gradually absorbed by the two permanent teeth. At fifteen months old the two permanent teeth are well up, and are very distinct from the temporary set, as they are broader and whiter. As at this time shearing generally commences, sheep at this age are sometimes called *shearlings* or *two-teeth*. At a year and ten months the second pair of front teeth appear well up, having come through the gums at any time after a year and six months. Consequently, if the second shearing is done early in spring, there are two pairs well up and the animal is called a *four-teeth*. The next pair come up quicker, and by two years and three months there are three pairs well up ; therefore, in late shearing, there are *six teeth*. More quickly the fourth and last pair come up, so that a little before three years are reached the jaw is full and the sheep is called *full-mouthed*. The corner teeth do not always come up level until the expiration of several months. After this, as time goes on, the teeth wear down, and being narrower at the base, they appear to get wider apart.

The *molars* come through at the following periods :—At a month three temporary molars are well up on either side of the upper and lower jaw ; at three months the first permanent molar is cut, appearing behind the hindmost temporary molar, and owing to the three temporary molars being afterwards replaced by three permanent molars, is called the *fourth molar*. At nine months the second permanent molar appears behind the first. This is called the *fifth molar*. At eighteen months the sixth and rearmost permanent molar is cut, and soon after this the first two temporary molars are replaced by two permanent molars, and the third molar is reduced to a shell covering the top of the permanent tooth: before two years the latter will have come through, but will not yet be level with the others. When this tooth becomes level with the others the sheep is over two years. The dentition of the molars is more regular than that of the incisors, so where absolute accuracy is required they afford the best evidence of the age.

Nomenclature.—The names by which sheep are known vary much in accordance with the locality. In the South of England, where sheep are now brought to the butcher when very young, the terms used to distinguish older sheep are becoming more or less obsolete; but in Scotland and in hill districts generally, where the feeding is not so much forced, the distinguishing terms are still required as much as ever. When first born the term *lamb* is universally used, the mother and lamb being collectively called a *couple*; if twin lambs, *double couples*. From weaning to first shearing they are called *hoggs* or *tegs*, tegs being a corruption of the word tags, as before the wool is clipped the locks taper to a point or tag; when once the wool is clipped the locks show a blunt end. In the South-Eastern counties there is an exception, the shearlings being sometimes called *tegs*. The terms *hoggets* and *hoggerels* are used in some localities.

The males are distinguished by the prefix *he* and the females *she*, though the males are sometimes called *wether tegs*. If the male lamb is not castrated he is called a *ram lamb* until he is shorn, when he becomes a *shearling* or *two-tooth ram*, and afterwards he is called *two-shear* or *three-shear* as each year goes by. A ram showing one testicle is called a *rig*. When the castrated male lambs are first shorn they are called *wethers, shear hoggs* (sometimes pronounced *sharrig*), or *wether hoggs*. In the South few are kept longer than as shear hoggs, but mountain sheep are kept on as *three-tooth wethers* and *four-tooth wethers*. Female sheep are called *theaves, gimmers, chilver*, or *two-tooths* at the first shearing. After that, as they are usually used for breeding, the terms *two-shear* and *three-shear*, or *four-tooth, six-tooth*, and *full-mouthed* are employed in accordance with their age. A ewe which does not breed is called *guest* or *eild*; when not in milk a *yeld ewe*; when withdrawn from the flock a *draft ewe*. The breeding ewes are called *stock ewes*. Other terms are occasionally used by breeders of and dealers in sheep, but those given are most often applied.

CHAPTER XI.

THE RAM AND EWE.

The Choice of a Ram.—The points which should be observed in buying sheep, already detailed, hold good with respect to the ram. A few other features should be looked for as the ram is the cheapest means of improving the flock. It is well said that the ram is "half the flock." This does not imply that the ewes should not be carefully selected, as perfection cannot be approached if the ewes are seriously faulty; but, fortunately, the influence of the ram is very great. In the first place, a ram influences the offspring of from fifty to one hundred ewes yearly, whereas the ewe only affects her own lambs. It is highly important that the ram should come from a stock which has been carefully bred and selected for many generations. A ram with a good pedigree possesses a fixed type which he imparts to the offspring. The greatest effect of a well-bred ram is apparent when he is mated with poorly-bred ewes: he possesses greater potency over them than he does on ewes which have been well bred for a long time, as the latter have also obtained fixity of type, and he does not make the same impression on them. From this it must not be implied that an inferior ram can be used on high-class ewes without doing harm, as the male is naturally prepotent over the ewe; but merely that the extraordinary prepotency of the well-bred ram over the poorly-bred one is shown to an exceptional degree. This is well, as it enables a farmer to improve his flock quickly, and at small expense. By judicious selection of his ewes, withdrawing from time to time those which are of inferior type, the whole flock may be raised to a high standard in a few years.

In selecting a ram for mating with poorly-bred ewes, the first point to look for is quality. If the ewes are small, size must be looked for also, but a squarely framed ram, with every characteristic of early maturity, must be sought for. A few pounds extra laid out on a good ram is money well spent. While, however, recommending that high-class rams should be used, it is not necessary for a farmer who intends improving his flock for purely farm purposes to purchase the most expensive rams, a few of which are found in most show flocks : these involve an exceptional outlay as they are required by others who intend to compete for prizes, and they are a speciality for which a higher price is paid than a farmer ought to expend. On the other hand, the inferior sheep of a flock are not likely to do so much good as those which possess some slight fault which precludes them from the show-ring but will not be seriously apparent in an ordinary flock. Cross-bred sheep do not breed so reliably to type as those which have been kept pure for a long time. The ram should show vigour, and a masculine character. This is usually indicated by a strong, though not necessarily coarse, head and a thick neck. The skin is an important feature, as poorly-bred ewes are, as a rule, inferior in this respect. The skin should be soft, pinky and bright in most breeds. From such a skin wool will generally grow closely and fine. In breeds descended more or less directly from horned breeds, signs of horns should be avoided. A ram should stand well on his legs and move freely.

As a rule a shearling ram is preferred. However, the Hampshire breeders, who have been very skilful in developing their sheep, prefer a ram lamb ; and they attach much importance to this, as they have found that the use of ram lambs has had a great effect in developing early maturity, for which the breed is so justly famous. A ram lamb should not be mated with more than fifty ewes, and he then will be fit to serve as many as eighty in the next season. If overworked, his successful career will be of short duration ; and if a high price is paid for a good ram lamb, at least two season's work should be got out of him. A specially good ram may be used for several seasons—in fact, so long as he is active. After two years, however, there is risk of inbreeding, as his offspring may have been brought into the flock. Where the ram is used for more than two years the ewes should be selected so that mating with his own blood may be avoided. Sometimes, however, inbreeding to a slight extent may be needed to secure fixity of type, or to tone down coarseness, which may have been brought about by injudicious mating or by the soil—some soils having a tendency to make the sheep run coarse.

Number of Ewes.—A shearling ram will take eighty ewes. It is important to purchase a ram in good health, and a ram with foot-rot should be avoided, as it will convey the disease to the flock with which it is put. Should a ram affected with foot-rot be purchased, it should be isolated until all traces of disease have disappeared. If very lame he may be incapable of service for the mere want of locomotion.

A ram requires to be in good condition at the time of service, but should not be over fat. When got up for sale, rams are often soft, and rapidly lose flesh when turned on to poorer food and put to service ; it is, therefore, advisable to get the sheep inured to harder rations by giving them less fat-making foods, but keeping up their vigour with flesh-making foods, and allowing them plenty of exercise. Rams which are strange to each other often fight vigorously at first, and should be watched. Some rams acquire the knack of breaking an opponent's neck. We know of a case where one ram destroyed three others which were consecutively put with him. When the ram is put to service, his brisket should be rubbed with a mixture of oil and ochre, to show which ewes he has leaped. The colour of the ochre should be changed in a month, so that those which are not in-lamb but come over again may be detected. The time of lambing will then be more accurately ascertained, and as it is often advisable to separate those which are due to lamb early from those which come later, an easy means of distinguishing them is thus obtained. After the rutting season the ram should be taken from the ewes, and may be kept with the wether tegs, where he generally finds a sufficiently good diet. It is not advisable to let him get too poor. He should be kept well on his feet, and his feet should be pared, so as to keep them in good shape.

The Ewe.—The male generally influences the outward form of the offspring, and, it is popularly held, the female the constitution. It is, therefore, important to breed from ewes of vigorous type. It is not usual to breed from ewes until they are two years old. Often when the stock of sheep in the country is small, lambs are put to the ram so as to produce young at a year old, but this is rarely attended with great success. The difficulty of lambing, and the short supply of milk they are able to produce, together with the dwarfing of size of the ewe itself, are generally sufficient to prevent those who make the experiment from repeating it. Some of the most successful Southdown breeders have been following the practice, and are well satisfied. It is noticeable, however, that those *on* the South Downs do not do it, and that it is usually done by those on stronger land, where the sheep have a tendency to grow rather big.

The ewe flock, when not directly bought in, should be maintained by selecting the best theaves each year, and putting them into it in the place of those which, from various causes, are drafted from it. It is not customary to run the ewe lambs on expensive lines when it is intended to put them into the flock. A lamb while developing into an ewe has to be kept two years, and if expensively fed throughout that time, is a costly animal. Economy in management must therefore be exercised, but the extreme of insufficient feeding should be avoided. It is advantageous for the ewe to be well grown ; consequently, although unnecessary expense is wasteful, she should be kept in a thriving condition. By selecting a draft of the best lambs the flock is improved yearly, especially when the ewes are put to a high-class ram. An endeavour should be made to obtain similarity of type and features, as sheep, whether sold in a large flock or in small pens of five, always realise a better price when they are well matched in colour and shape of the heads, in size, and in quality of the wool. Quality is the English sheep-farmer's watchword, as it is only by producing meat of better quality than the foreigner that he can reap the advantage which the soil, climate, and better farming of England afford. Let the aim therefore be quality. This is more difficult to get than size. Too great size, especially in breeds whose special value is in their mutton, is a mistake, as coarse joints are not wanted. Quality is, however, compatible with size, but quality must stand first. Select the ewes with this view. Refuse those which are coarse, gaunt, and narrow in the forequarters.

Culling Ewes.—After each weaning the flock should be over-hauled, so that those no longer worth keeping in it may be taken out to make room for the draft of young ones. The first to go should be the barren or guest ewes ; then those which have had diseases of the udder, or are abnormally deficient in their supply of milk. A short supply of milk is not uncommon in ewes with their first lamb, so too much notice should not be taken in the first year. Those which had an inversion of the womb should not be bred from again. As ewes have to " cut " or graze much of their food, often from bare pastures, or following other sheep on roots, it is necessary that their front teeth should be sound. On stony land the teeth are frequently broken off when they are com-paratively new, and they gradually wear away under any circum-stances, so that from the time the teeth are fully developed their grazing powers are lessened. When they can no longer graze sufficiently for their proper sustenance, it is of no use to keep them in the flock, as the extra labour of producing a lamb brings them to the point of starvation, and either the ewe or the lamb is bound to suffer. The ewes withdrawn from the flock should be sold or fattened off as quickly as possible, according to the food at disposal.

CHAPTER XII.

THE GREEN FOOD SUPPLY.

The simplest system of sheep-keeping is that by which the animals are fed on grass throughout the year. Here the supply of food is dependent on the amount of grass grown, and varies with the productiveness of the season. The system most requiring skill and forethought is where the sheep are kept entirely on crops of arable land, while the mixture of grass and arable land crops is midway between the two. The crops which are most serviceable on arable land may be divided accordingly as they are suitable for particular times of the year. It is usual to divide the sheep year into two portions, one commencing at the beginning of November and lasting until the end of April, known as the winter season, and the other occupying the remaining portion of the year, known as the summer season. The winter is the season of root crops, the summer of grass, clover, and other green crops. The same kind of roots are not so valuable at all periods of the winter season, and a succession of crops suitable must be arranged. By dividing the year into the following periods, the crops which are most reliable, and are best suited in ordinary seasons, are readily seen, though, of course, they overlap to some small extent.

November to February.	March to May.	May to August.	September and October.
Swede turnips	Mangels	Clover and seeds	White turnips
White turnips	Kale	Grass	Drumhead cab-bages
Drumhead cab-bages	Winter greens	Trifolium	Late rape
Kohl-rabi	Early catch crops, such as : Tares or vetches, winter rye, winter barley, winter oats	Early cabbages	Stubbles
Grass		Early rape	Grass
Ensilage		Winter and sum-mer vetches	Young seeds
Late-sown turnips run to top		Sainfoin	Hay
Hay	Water meadows	Lucerne	Kale, sown early
	Late-sown turnips run to top	Second feed of kale	Kohl-rabi, sown early and close-ly, but not singled
	Rape		Sainfoin
	Hay		Mustard

Swedes.—Swede turnips should not be fed before November, as, although they may have acquired size, they are not ripe.

Unripeness causes scour, and many sheep are lost by being put on to them too early. The food contained in them is also not in a form which can be fully utilised. From November to March is the season of the swede. After March they become dry and pithy, much of their feeding value having been lost.

Kohl-rabi.—Kohl-rabi cover much the same period as the swede, the Early Small Top variety being excellent early winter feed ; the Late Big Top, being hardy, is valuable at a later period, as it withstands fairly severe frost. The modern Big Top late kohl-rabi is of better quality than used thirty or more years ago, has thinner skin, and is easier to gnaw ; but it cannot stand the same severity of frost. When planted, care should be exercised to obtain the variety suitable for the season at which the crop will be required.

Mangels.—Mangels are essentially spring food, although they are frequently fed at an earlier period. It is wasteful to feed them early, as the food is not wholly digestible. The greatest value is found in them from March to June, if they can be spared to be kept so long ; the greater part of the food in them is then made use of by the animals. When used for male sheep in the spring, the yellow shoots should be stripped off, as they tend to produce urinary troubles by forming crystals in the passage from the bladder, which check the flow of urine, and cause inflammation of the bladder, often resulting in death. Even without the leaves, there appears to be a tendency for these crystals to form, but to a far less degree. Ewes, being possessed of a bigger passage, rarely suffer.

Cabbages.—Cabbages supply a succession of food from June to Christmas, if early varieties are transplanted in autumn, and later varieties of Oxheart type are transplanted in spring. These will be available until September, when the Drumheads, also transplanted in spring, or drilled and set out very early, will be available until the end of the year ; but they cannot be relied upon longer, as when thoroughly ripened they are susceptible to destruction by frost. On the Suffolk coast the weather is generally so mild that cabbages are relied upon to produce lamb food in March.

Probably the weakest spot in sheep management is the small reliance placed on cabbages. They make by far the best " root " food given to sheep ; and whether used in a late summer or autumn are the most dependable food in droughty periods and after, and in point of production and cost of placing before animals, are cheaper than any crop approaching them in value as sheep food.

Kale.—Kale, such as the Thousand-head, Shepherd's, and other hardy varieties, have a special value, coming to their best between the early part of March and May, supplying rich, succulent food during the hard pinch between roots and grass. Shepherd's, when planted early, produces a heavy crop early in autumn and winter, owing to its exceptionally thick leaves and rapid growth, but does still better when left until spring for a greater quantity of side shoots to push out. The tardiness with which Thousand-headed kale comes on at first prevents many from using it, but it crops so well when it has been on the land a year that it is worthy of a more extended cultivation. However, it can be sown thickly, and be fed in the autumn before it throws its many sprouts which appear in spring, and give it its name. The advantage of both these kales is that if a cutting is taken in April, another crop will be available in July.

Rape.—Rape is of two types, Giant and Dwarf. The Giant produces a single crop, and, if sown in May or June, is fit to feed in August and September, when it attains great size. The Dwarf kind may be sown from March to July, and will produce several crops before it is necessary to destroy it. March-sown rape will often produce good feed in mid-June, another in September, and valuable green food in spring. Lambs should not be put on it a second time unless it has been well frosted, as it becomes " sour," and then causes scour. As a rule it is sown at the same time as white turnips, and then gives an autumn and a spring crop. It is valuable as a catch crop taken off land from which an early summer crop has been taken. It then makes excellent lamb food in spring.

White Turnips.—White turnips are available from August to December. Those sown as a catch crop in July and August are useful in spring, when the tops provide good lamb food. Quickly maturing varieties like Stratton Green and Six-weeks are valuable in seasons when the main crop of roots has failed. Sown in July and early August, they make useful autumn feed.

Greens.—Winter greens, such as Savoys and Brussels Sprouts, are very hardy, and stand the severest weather.

Rye, Barley, &c.—Rye, winter barley, winter oats, winter vetches, and trifolium are essentially autumn-sown catch crops, available at the critical period after the root crops are consumed. Rye is the first to be fit for feeding; vetches and trifolium come a little later. A small piece of winter barley is valuable on all farms where lambs are raised, as it is one of the best antidotes to scour, the great scourge of the lamb-breeder.

Mustard.—Mustard sown as early in August as possible makes welcome food in October, and is much liked for "flushing" ewes. It can be sown at other periods from late spring onwards, to be fed within two months.

Ensilage.—Ensilage is of great value to the sheep-farmer, and silage is especially valuable as food for ewes, as they eat it readily, thrive well, and it produces a good flow of milk. It is a convenient and most useful fodder in long-continued frosts, when roots get frozen into the ground, and are difficult to obtain in sufficient quantity. The comparative warmth of silage is also a matter of great consideration. Frozen roots are indigestible, and sheep thrive badly on them, as the goodness they contain is expended in maintaining animal heat within the body, and little goes towards improving the condition of the sheep. We have a high opinion of the value of silage, and strongly recommend its more general use. An expensive silo is not necessary, a field heap being economical in all respects except outside waste. Clover and other seeds make good silage, and if made in the field where grown, the cost of carting to the silo is small.

Sainfoin.—Sainfoin, clover, lucerne, grass seeds, and similar crops grown as short leys are the mainstay of the sheep-keeper on arable land during summer, and in one way or another they may be relied upon from May to November. They are also the chief source of hay for winter feeding.

Stubbles of corn crops afford sweet fresh food, and are specially valuable when they contain strong, young seeds.

Pasture affords keep at almost all seasons.

Water Meadows have a special value as affording a bite of fresh grass as early as the first of April.

Catch Crops.—Catch crops are of so much importance to the sheep farmer that more than a mere notice of their value should be given. They may be regarded as being specially valuable on two occasions—in time of drought, to augment the limited amount of sheep food during the coming months ; and in ordinary seasons, for utilising the land when it would otherwise be lying idle and losing its manurial constituents. Crops which are suitable for sowing to produce keep within the shortest time are mustard and stubble turnips. In dry seasons the land earliest cleared of corn or other summer crop should be brought into a good tilth, and be sown with one of these to produce early autumn keep. It should never be forgotten that every day is of value, and that no time should be wasted in getting the land sown, whatever catch crop is taken. Hardier varieties of turnip, kale, and rape, sown as catch crops, will produce excellent feed in spring, and be particularly valuable for lambs. The autumn-sown catch crops, suitable for feeding in late spring, such as large winter barley, oats, vetches, and *Trifolium incarnatum*, should be sown as early as possible. The cultivation for all but the trifolium consists of nothing more than ploughing, sowing, and harrowing, a very simple seed-bed being sufficient. Trifolium thrives better when the land is not ploughed, but merely harrowed.

CHAPTER XIII.

Concentrated Feeding Stuffs.

The most valuable feeding stuffs are linseed cake, cotton cake, Soya bean cake (with oil reduced to 1 per cent., as the oil has dangerous properties), special feeding cakes and mixtures sold for sheep food, wheat, barley, oats, peas, beans, maize, lentils, pea husk, malt, malt culms, bran, rice, and linseed. Fenugreek, ginger, and other spices are employed to give an aromatic and enticing flavour to other foods. The terms *cake* and *corn* are used somewhat indiscriminately in some districts, farmers frequently saying that they are giving their sheep so much " cake," when a portion of it is " corn," or so much " corn " when some of it is " cake." What they mean to imply is that they are giving a certain quantity of rich concentrated food.

Linseed cake is the best individual concentrated food. It contains flesh-forming and fat-forming constituents in good proportions, and the oil acts beneficially on the bowels. The ease and safety with which it can be given makes it very popular with flockmasters, and this tends to keep the price somewhat higher than that of a mixture of other foods of the same value, which can be prepared by any one knowing the constituents of cake, and of the other articles on the market. But it can be given to dangerous excess.

Cotton cake is sold in two forms, known as " decorticated " and " undecorticated." The former is freed of husk, and in the latter the husk is left. The feeding value of the decorticated is much greater than that of the undecorticated ; it is richer in feeding constituents, and is not so astringent. It should be broken very small, as almost all samples are hard ; those containing hard, brown patches varying in size from a bean to a half-crown in circumference are particularly dangerous, as these pieces are almost indigestible, causing great irritation in the stomachs of old sheep, and frequently prove fatal to young sheep. These hard pieces must therefore be broken, and for this reason the cake should not be given in lumps larger than a bean. The astringent principle found in undecorticated cake makes it a suitable food in some forms of scour. All cakes should be broken small, but it is of the greatest importance in the case of decorticated cotton cake. As cotton cake contains a large proportion of flesh-formers, it should be given in mixture with foods containing more starch. *Special feeding cakes* are usually well compounded and safe foods. This is not always the case, and sometimes very inferior stuff is employed in their manufacture. Such foods should be bought on analysis, and care should be taken on delivery that they are sweet and in good condition.

Wheat, barley and oats are valuable foods, and at low prices are economically used as sheep foods. As a rule they are better for being mixed with peas, cotton cake, or other nitrogenous food. With all grain and starchy foods a small quantity of crushed linseed is of special value, as the oil greatly aids digestion and helps to maintain the health of the animals. By themselves they are found to be heating. *Peas, beans and lentils* are excellent sheep foods, helping to form a good proportion of lean meat, and making the sheep handle firmly. It is important that peas and beans be not used when they are new, as they are then indigestible, causing scour and other disturbances, and do not yield their full feeding powers. They are not considered old until the March following their harvesting, and it is better to let them remain a full year before being consumed. Peas, beans, and all corn and grain should be passed through a kibbling machine. It is advisable not to reduce them to a fine meal, but to crack or grittle them, as meal is liable to be blown out of the feeding troughs, or in wet weather to form a paste. Large quantities of unbroken corn swell sufficiently to distend the stomach unduly ; and the full value is rarely obtained when given whole. The cost of grittling is well repaid.

Maize and rice are starchy foods, well adapted for mixing with the more nitrogenous foods. *Malt culms, pea husk and bran* are valuable foods, and have a special utility for mixing with chaff or chop to make animals eat a larger quantity. Pea husk is very suitable food for lambs, and we know of no other food which keeps them in such good health. *Hay,* if not a highly concentrated food, is a valuable aid to the sheep-keeper, as in addition to a fair amount of feeding properties, it tends in no small degree to keep the animals in a thriving and healthy condition. Sainfoin, lucerne, broad clover, and " mixture " from temporary leys are, if anything, preferable to meadow hay, although the latter is very valuable. Sainfoin and broad clover are the best. It may be given in racks in ordinary condition, or in troughs chaffed.

CHAPTER XIV.

SECTIONS OF THE FLOCK.

Composition of the Flock.—The previous chapters have been devoted to matters which have an important bearing on sheep management, and they should be sufficient to enable the reader to follow with ease descriptions of systems of management which will be given in the subsequent chapters. The most important sections of sheep farming are included under the following headings :

> Management of Ewes.
> Management of Lambs.
> Management of Stores and Wethers.
> Management of Fattening Sheep.

These are subject to many subdivisions, and an endeavour will be made to lay before inexperienced sheep-keepers the work of the flockmaster from day to day and from season to season, according to the systems of farming which may be practised. As grass sheep-farming is not so complicated a calling as arable sheep-farming, attention will be mostly directed to the latter, especially as the proper management under this heading includes most of that under grass-farming.

Under ordinary circumstances a flock is composed of ewes, lambs, store and fattening tegs, a few rams, and wethers. The ewes are regarded as the scavenging portion of the flock, as, except at lambing time, they clear up behind other sheep, and are run as inexpensively as possible consistent with their healthy maintenance. Lambs which are to be fattened off as fat lambs, or tegs, are kept on the best and freshest food, and generally receive cake or corn in addition. Ewe lambs to go into the flock are not forced, but are kept on sweet fresh food. Where an endeavour is not made to get out sheep until they become wethers, they are not fed highly or expensively kept until some little time before they are fattened out. A chapter is devoted to the management of a show breeding flock, and it will be better not to deal with them here, as they form a distinct feature in sheep-keeping.

Working a Mixed Flock.—Perhaps the best way of showing how a flock is managed is to take an imaginary farm of mixed grass and arable, and show how a flock of 200 ewes and their offspring may be maintained on it. The size of the farm need not be definitely fixed, as on such a farm cattle are always kept, but sufficient grass and arable must be allotted from time to time to maintain the sheep at various seasons. Different soils possess varying powers of carrying sheep, as may be gathered from the fact that on some of the arable sheep farms, with a small proportion of grass and water meadow, and with a mixed cropping of corn and fodder crops in which the fodder crops embrace the larger acreage, from two to two and a-half sheep per acre over the whole acreage of the farm are sometimes permanently carried. This is particularly the case where an extensive system of catch cropping is adopted, whereas on the thinnest hill land and heath some acres have to be given up to one sheep. On a mixed farm in the Midland counties, with a small proportion of grass, but with the arable land fairly divided among corn and fodder crops, one to one and a-quarter sheep over the whole farm may be carried, though this number is found oftenest in winter, when they are on the roots, an extra number of tegs having been bought in for wintering. The best place to take up the treatment of a mixed flock is where the ewes have been deprived of their lambs, as then the whole of the breeding and management of the lamb until he is sold lies in the future. The stock is then complete also.

The season of *weaning* varies in accordance with the period at which the lambs are born, and is generally about four months from the time of birth. Where lambs are born very early, however (irrespective of show animals), it is customary to dispose of the earliest as Easter fat lamb, the sales continuing for some time. Dorset lambs, which are bred in the autumn as early as October, are sold in early winter. Dorset ewes may be put to the ram in May; and often breed twice in one year, being put to the ram a second time as soon as they come into season again. Hampshires and Oxfords are born in January, February, and March; Shropshires and Southdowns in February and March; and white-faced sheep in February, March and April, according as the grass is likely to come in. The *lambing season* is practically controlled by the season when suitable food can be obtained, and the time when grass will be ready for the lambs is the chief regulator. On arable land, however, the cropping can be arranged so as to be available at any time, and the time of grass need not be regarded so closely. On hill land and mountains the lambing time is sometimes deferred as late as June. For the purposes required here, as affecting sheep on a mixed farm, February and March may be taken as the lambing season; consequently, the weaning season may be put at June.

Head of Sheep at Different Seasons.—A certain loss is always experienced, and it usually varies from 3 to 5 per cent., but from abnormal circumstances may go far beyond this. However, in the estimate given here they are not taken into consideration, round numbers being adhered to.

Taking the flock of ewes as being 200 at the time of weaning, they may be estimated as having brought up 250 lambs, fifty of which have been sold as fat lamb. There should be fifty theaves ready to come into the flock. Lambs weaned at this time in the previous year should not be kept on to be fattened as wethers : therefore, in this case they are considered as having been sold, though there are instances where they might be run on to be fattened on the seeds during summer. The stock is, therefore, 200 ewes, 200 lambs, 50 theaves ; a total of 450.

Immediately after this, fifty ewes are culled from the breeding flock, and may be sold at once, or be fattened during summer. The chance of selling the ewes affords an opportunity of lowering the number to be kept through the summer, should the prospect of keep be poor. The food available consists of seeds or leys, vetches, and early cabbages. On a mixed farm it is usually advisable not to stock the pastures until after they have been mown.

The same number of sheep, less the fifty culls, should be kept through the autumn, and it depends on how well the tegs have thriven when they are fit to be sent to the butcher. As fat tegs, the earliest rarely go to market before the end of November, so the winter season is entered with the same number. During autumn the ewes will have found food on the stubbles and aftermaths of grass and leys, together with a little stale food behind the leys or cabbages and white turnips. The lambs at the same time may have had the first pick over the stubbles, grass, and leys, and the first gnaw at cabbages or white turnips, with a small quantity of cake or corn.

The winter season (from November to May) sees great alterations ; the tegs are gradually sold, and a new crop of lambs appear. Half the tegs may go before the lambs are born, and the other half, minus fifty she-tegs which are to be kept on as breeders, may be sold subsequently. This allows for thirty to go in each month from December to April inclusive. It will, however, depend very much on the quantity of cake they have received from weaning as to when they are fit to go out, and if it has been moderate, the whole 150 may go out between February and April inclusive. This has to be borne in mind when arranging the cropping.

More briefly put, there would be at the weaning season, 1914, 200 ewes, 200 lambs, and 50 theaves on the farm, making 450. However, 50 cull ewes would soon be sold, leaving 400. At the 1st of January, 1915, the same sheep remain, less 30 fat tegs sold, leaving 370.

During January, February and March the remaining tegs, less fifty ewe tegs to be retained for the flock, are sold, reducing the number of sheep to 250. Meanwhile 250 lambs are born, so that the total is 500 ; but fifty of these are sold fat in spring, bringing the total at weaning-time, 1915, to 450, as at the same period in

1914. This is regardless of sheep which may have died or have been purchased.

The Shepherd's Account Book.—It is important to keep a strict record of the sheep, as with fluctuating numbers there is no other way of proving loss. The shepherd should keep a note-book, in which he should record all sales, losses, and purchases ; he should also note the number taken into each field, and the number removed from the field after the food is consumed, so as to make a check by which he can prove the loss by straying or stealing of any sheep in his charge. Moreover, a daily count should be insisted upon.

The farmer should check his book, and post it in his sheep account. He will find elsewhere a useful form in which to record the sales and purchases of sheep, also the births and deaths. But, usually, farm account books provide methods which accord with the general principles running through them.

Sheep Required to Feed off the Root Crops.—As a rough estimate of the carrying powers of an acre of cropping, it may be taken that a teg having an allowance of corn, eats its own weight of swede turnips each week, and an extra sixth more of white turnips. An ewe will eat at least half as much more. Thus, taking the average of the tegs to be eight stones, each sheep will eat 1 cwt. per week ; and each ewe 1½ cwt. A twenty-ton crop of roots, which gives 400 cwt., will, therefore, keep 200 tegs for a fortnight, and 200 ewes three-quarters of a fortnight.

THE SHEPHERD'S ACCOUNT BOOK.

Sheep Purchased.

Date.	From whom.	Class of sheep.	No.	Price.	When received	When paid.	£	s.	d.
1914.									
Jan. 1	In stock	Ewes	198	55s.	——	Valuation	544	10	0
Jan. 1	In stock	Tegs	146	44s.	——	Valuation	321	4	0
Jan. 1	In stock	Rams	3	80s.	——	Valuation	12	0	0
Feb. 12	Bred	Lambs	140	——	——	——	——		
Mar. 1	Bred	Lambs	137	——	——	——	——		
July 28	R. Jones	Wethers	50	37s.	July 28	July 28	92	10	0
Aug. 7	R. Smith	Ram	1	10gns.	Aug. 17	Aug. 17	10	10	0

Sheep Sold.

Date.	To whom.	Class of sheep.	No.	Price.	When delivered	When paid.	£	s.	d.
1914.									
Jan. 14	By auction	Tegs	20	45s., less 10s.	Jan. 14	Jan. 18	44	10	0
Jan. 21	W. Hill	Tegs	25	50s.	Jan. 24	Jan. 28	62	10	0
Jan. 23	Died (S. Brown)	Ewes	3	5s.	Jan. 10, 21, 23	Feb. 17	0	15	0
Jan. 27	Died (S. Brown)	Ewe	1	7s.	Jan. 28	Feb. 17	0	7	0
Feb. 5	W. Hill	Tegs	15	47s.	Feb. 6	Feb. 8	35	5	0

In arranging for the feed until May, the eating capacities of the earlier lambs have to be taken into consideration. The earliness with which they are born of course greatly controls the quantity of food they will eat ; and this food will rather be an equivalent to roots, than roots themselves, as they require soft succulent food, such as turnip tops, rape, or kale. All food, however, is best estimated on the equivalent of swedes, or other standard crops. Mangels, kohl-rabi and cabbages (as long as they last into winter) may be regarded as the equivalent of swedes. Sprouting crops are more difficult to estimate, as the severity of the winter and the backwardness of spring growth affect them. The feeding power of a kale crop may far exceed that of roots even in winter time, and to a greater extent in spring ; but this only relates to that planted early in the previous spring. Late planted kale and rape give small return in winter ; but when kale gets its spring shoots in April, it crops very heavily.

It must always be borne in mind that there is greater likelihood of decrease than increase in the supply of food, as the risk of injury from frost, mildew, and other destroying powers has to be run ; therefore, it is unsafe not to make the provision per head in excess of bare estimates. The strain on the food will be heavy in the early spring months, when the 200 ewes, and a portion (say half) of the tegs are still in hand, and the lambs are beginning to feed. All sections of the flock require to be well fed at this season : the ewes to supply milk for the young lambs, so that they may not receive a check in growth, the tegs so that they may come rapidly into condition for killing.

The food available during March and April consists of a portion of the swede crop not consumed, mangels, autumn-sown catch crops (except tares), rape, and kale. In May the leys and grasses may be looked to for help ; trifolium and rape, together with autumn-sown catch crops, including tares, should also be available. The fat tegs should have been sold ; but the place of the latter portion sold is more than taken by the theaves and young lambs, which by this time have acquired big appetites. Where the land is largely composed of pasture, or where grasses and clovers are grown in rotation, the flockmaster can generally feel secure of feed throughout May.

If the food of the farm shows at any time that there will be more than is required to carry such a flock as has been described, the farmer has a chance of buying in more, if the prices look favourable. It is, however, foolish to buy sheep which show no chance of profit. It is better to take in others at agistment, at a small price per week. Failing this, it is wiser to plough in a portion of the crop as a green manuring. By doing this, the land gets the advantage of earlier working, which is always beneficial.

CHAPTER XV.

A Year with the Ewes.

Having given a brief description of the position of each section of the flock at various seasons of the year, the details of management can now be gone into. The ewes are the first section to deal with, as they are the most permanent portion of the flock, and as the lambs and future sheep are dependent on them, it is most convenient to treat of them.

The Ewe at Weaning Time.—Taking up the subject of the ewe at the time the lamb is weaned, the first thing to see to is that she does not suffer from weaning. Attention should be given to the udder, to see that it is not unduly inflamed or distended. A point is always gained when the ewes are shorn before weaning, as then a casual observation of the udder reveals its condition. When the wool is overhanging, it is necessary to catch the sheep and handle the udder. If milk continues to form in large quantities, the udder should be drawn a little to ease the distension, but it is not necessary to empty it. A dose of Epsom salts will help to relieve it, and may be used without fear of injury. Succulent food should be withheld, and the diet should be somewhat scanty and poor. In case inflammation sets up, the same treatment must be more rigorously adhered to, and the udder should be well rubbed with lard. If care is taken at the outset, serious complications rarely arise. No time should be lost in getting the flock overhauled ; those not suitable for breeding from again should be culled in accordance with the instructions given in another chapter. The culls, if not sold at once, should be put on to good feed, and, if necessary, be given cake to fatten them, as no good end is gained by keeping them on the farm longer than there is absolute need of. They may be sold off as they become fit for the butcher, or all at once, as appears best. No other special treatment is necessary. In some of the large breeding districts the ewes are sold off as soon as they have produced and reared a lamb after becoming full-mouthed, and are then bought up for breeding in other districts, where the custom prevails of buying in such ewes with a view to letting them breed once more, and then fattening both the ewes and the lambs simultaneously—a practice which is often attended with considerable profit.

The ewes having been drawn out, should be branded with a pitch brand. For the better identification of the sheep, it is advisable to mark all of them in special parts, according to their age and sex. Various systems are adopted, but nothing is simpler to remember than the following, which embraces the ordinary sections of the flock :—

> Ram tegs, on the off shoulder.
> Ewe tegs, on the near shoulder.
> Theaves, in the middle of back on near side.
> Wethers, in the middle of back on off side.
> Two-shear ewes, on hip on near side.
> Three-shear ewes, above tail.
> Four-shear ewes, on hip on off side.

A special effort should be made to get the feet sound, as effects of the yarding are often in evidence where proper care is not taken. This is one of the most leisurely periods of the shepherd's year, and advantage should therefore be taken of it in this way ; also in mending hurdles, or getting forward with anything which has been allowed to drop in arrear during the previous busy season.

Treatment of Ewes before and during Gestation.—The breeding flock should be got into good condition to receive the ram, as then a greater fall of lambs may be expected in the following spring. Ewes falling in condition do not come into season so early, and are less likely to produce twins. If the ewes do not come into season at the desired time, a little heating corn, especially barley, will help to bring them on. Barley and wheat stubbles are useful in this respect. After being put to the ram, the ewes may be kept moderately well ; it is best not to allow them to become fat. Grass is a very economical food during the autumn and early winter, but the ewes may be put on the arable land at night to clear up food which other sections of the flock have left behind.

When the wool gets long enough to carry the dip, ewes should be dipped in one of the recognised dips, otherwise they will be seriously troubled by vermin, and in spring the lambs will catch the pests from them, much to their detriment. Dipping is now compulsory at certain seasons. Ewes go with young about twenty-one weeks, but this varies slightly, as does the period of gestation in most animals. During gestation the ewes should be allowed plenty of exercise, and this advantage is gained when they roam on broad pastures. Grass land is eminently suited to ewes in lamb, but when food gets short, especially during the last few weeks before lambing, they should receive extra food. The idea commonly prevails that turnips exert a baneful influence on the ewe and the lamb she carries. That ewes fed exclusively on turnips do give trouble at lambing, and often cast their lambs prematurely, there can be no doubt, but it is less from the presence of injurious matter contained in the roots than from the want of nutritive matter. The ewe for a few weeks before lambing has to build up the lamb within her, and this is a great strain on her. If she is in low condition and is supplied with insufficient nutritive food, Nature revolts, and either the ewe or the lamb suffers. The lamb most often suffers, and is expelled before time, Nature taking this course to save the ewe. The same result is obtained if ewes are kept too long on poor grass. We had experience of this some years ago, when, owing to a shortness of keep, we sent our ewes to graze at a distance. They stayed too long, and our losses at lambing were the heaviest we ever experienced. Our district being a strictly arable one, ewes rarely went on grass, yet by giving a plentiful supply of dry nutritive food with the roots, they were as prolific as in any other district. The fact is, turnips contain little more than sugar and water, and a ewe cannot support herself and build up a lamb on such a thin syrup ; it is not natural. She requires a more complex food, containing that which will supply the means of building up bone, muscle, and other parts of the lamb, without unduly exhausting herself. This must be borne in mind, and as the greatest strain comes when parturition approaches, care must be taken to give the ewe good food. At the same time the ewe should not be fat. She should be vigorous and muscularly strong. Dry food, such as hay, pea-haulm, or straw-chaff, with malt culms or other nitrogenous food, is well suited for this. A small quantity of peas, decorticated cotton cake or similar concentrated food may be given with much advantage, if coarser food cannot be conveniently spared. Corn of a starchy nature, such as maize, is not so valuable. It is a mistake to delay giving the ewes dry, bulky food until close upon lambing time. Hay or other dry food should be given in early autumn, when many cases of premature lambing would be prevented.

Care must be taken that as lambing time approaches ewes are not chased by dogs, or in other ways made to hurry themselves unduly. Ewes which are close-folded should be taken out for gentle exercise for a mile or two daily, special care being taken that they are not hurried through gateways, as the concussion is injurious to their young. Ewes—in fact, all sheep—should be approached gently, so as not to frighten them. Strange dogs should be kept well to heel. When ewes are well gone in lamb, they must not, if foot-rot is present, be turned to dress their feet ; but the feet must be lifted carefully and dressed or pared if they get badly grown. On those farms where permanent lambing folds are not found, it is necessary to provide proper shelter for the ewes and lambs during the lambing season. This should be in readiness before the most forward ewes are expected to lamb. Where temporary lambing pens have to be put up, they should be placed in the most convenient position with regard to food, so as to avoid unnecessary carting. This should be decided upon long before, and to provide straw for litter a stack of corn should be built near, so that when threshed the straw may be available. There is no better form of temporary lambing pen than that adopted on the Wiltshire and Hampshire Downs. Except on small or very compact farms such yards are preferable to a permanent one, as the carting of bulky material to and fro, and the expense of getting the manure made in the yards on to the land subsequently, is avoided. They well repay the trouble of erection.

The Lambing Yard.—Before describing the method of constructing a yard, it is advisable to point out what divisions of an ordinary lambing flock are usually found necessary. In dealing with a ram-breeding flock, many subdivisions of the lambs, according to sex and age, are necessary. The ewes require to be handled to see which are likely to lamb earliest, which is generally indicated by the udder, though young ewes frequently do not show much sign of milk until just previously to lambing. Those likely to lamb at once should be separated from those which will not lamb for some time ; drafts of the latter should be put with the former from time to time. Those which will lamb early should be yarded at night, but it is advisable to keep the others out of the yard as long as possible, because foot-rot is likely to be contracted in the yards, as the feet soften and become foul when kept on dirty bedding.

Around the ewe yard, a number of small pens, a hurdle square, should be fixed, so that, as lambing becomes imminent, the ewe may be separated. A pen should also be provided for the ewes which have lambed, and in large flocks additional yards are required, so that, as the lambs become stronger they may be taken out into the fields during the day. The divisions are, therefore, made in accordance with the size of the flock in the first instance, and the age of the lambs subsequently.

It is a good plan to build the yard so that the straw stack is in the middle, to facilitate rebedding. The method of making a yard as carried out in Wiltshire and the neighbouring counties is as follows :—Rows of hurdles are set up to form the outsides of the pens ; about 4 ft. inside these, posts about 6 ft. or 7 ft. in length are driven into the ground at intervals of about 10 ft. Deal battens about 3 in. by 2 in. are nailed from post to post to afford support to the roof. Outside the row of hurdles, but close to it, another row is laid on the ground, and on this a layer of straw is placed ; these are then lifted up so as to stand parallel with the

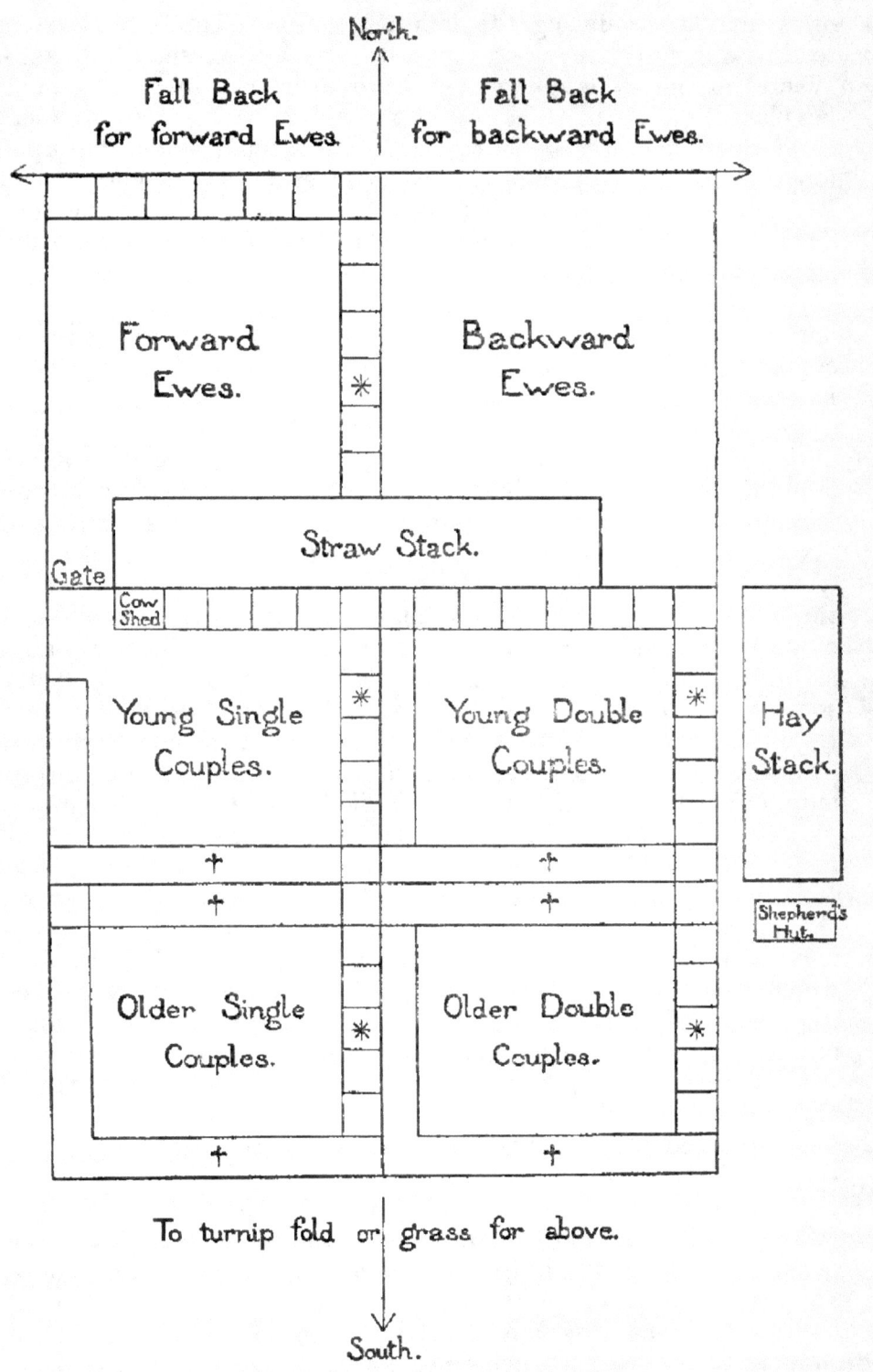

North.

Fall Back for forward Ewes.

Fall Back for backward Ewes.

Forward Ewes.

Backward Ewes.

Straw Stack.

Gate

Cow Shed

Young Single Couples.

Young Double Couples.

Hay Stack.

Shepherd's Hut.

Older Single Couples.

Older Double Couples.

To turnip fold or grass for above.

South.

GROUND PLAN OF FIELD LAMBING FOLD.
* EWES AND LAMB PENS. † SHELTER ALONG SIDE OF THE FOLD.

first row, being held in position by stakes. In doing this, the straw becomes held up between the double rows, and effectually blocks out the wind. Hurdles are then laid from the top of the back row to the line of battens to form a roof. A covering of straw is placed on these, and on this another layer of hurdles is laid and secured. A substantial roof is thus obtained in a very short time.

This shelter should be placed round all the pens, and a division to contain each ewe and lamb for a few hours after lambing should be provided by making partitions at each hurdle's length along the north and east sides. A line of hurdles placed at right-angles to the row will do this, and the front of the division can be closed by another hurdle. Each division will then be a hurdle square. A few only of these divisions need be provided for the older lambs, where they are useful in times of sickness. When the lambing season advances and the lambs require less shelter, the straw from the walls and roof is utilised, the destruction of the yard being brought about gradually. In other districts other methods of protecting the lambs are adopted, but the one next most serviceable to that described is that used on the large farms in the eastern counties, where, instead of stuffing the hurdles in the way described, two rows are set up about 3 ft. apart, parallel to each other, and at right-angles to these, other hurdles are set to form small pens. The space between the double row is filled in with straw, and to afford shelter, poles are laid on the dividing hurdles reaching half-way or more along them. The straw is spread over the poles, and formed into a roof which is roughly thatched. The principle of the protection is thus a long narrow stack, with the eaves carried over to a considerable degree. The straw used for protection becomes available for litter subsequently. Rows of bushes or wood are occasionally set up to break the force of the wind. The use of sheltering cloths attachable to hurdles has come into vogue, and they form an efficient and easily erected shelter, whether in the lambing pen or in the open field. In arranging the yard the ewes which have not lambed may be kept on the north side of the straw stack, and the young lambs placed on the south side. The straw stack is best built long and narrow. A haystack should have been built near, and the shepherd requires a portable hut in which to sleep and to keep medicine and corn. When the ewes are short of milk, a cow kept in the yard is useful ; in fact, the ewe's failure must be made good by the cow. It is well to point out, however, that cow's milk is often objected to by shepherds. Cow's milk rarely injures, though by giving the milk out of bottles rarely cleansed, covered inside and out by putrid milk, doubtless instances of harm frequently arise.

Ewes require a good " fall back " or space on which to rove. It is not always convenient to have this attached to the lambing pen, but it should be arranged that where they feed, even if a fresh piece is given daily, the hurdles are not kept within too close quarters. It is important to place the fold on a dry, solid field, an old ley being very suitable; if on a hill-side facing the south, so much the better. Subdivisions may be added. A fair number of subdivisions make the shepherding easier, as the sheep are more quickly handled.

The Shepherd's Requirements in the Lambing Pen.—A soldier cannot take the field unarmed, and a shepherd equally requires appliances, lotions, etc., to help him battle with the difficulties before him. The cow will supply the milk necessary for the lambs, which otherwise would not get sufficient. The cow should be neither too freshly calved nor too stale. The hut should be provided with a small stove to supply warmth, care being taken that the fumes can escape readily, as many shepherds have been suffocated by closing ventilating apertures on cold nights. The stove will heat the milk, warm the necessary water, fry the shepherd's rasher, and warm his billy, whatever it may contain. The vessel in which the milk is boiled should be made on the glue-pot system—a smaller one inside a larger one containing water to keep the milk from burning.

Among other things which he will have to provide are a ball each of string and stout tape. In those cases where it is desired that the string shall not slip, tarred cords are convenient. These are required for securing to the legs of lambs during malpresentations, for binding the side of the shape in cases of eversion, and for many other purposes during the course of the lambing season. Ruddle or ochre of several colours is required to make distinguishing marks. In pedigree flocks ear-markers may be used with advantage, to prevent ultimate confusion.

The shepherd will also need one or more drenching horns or bottles (an old sauce bottle with strong, long neck is very convenient); a cordial for chilled or weakly lambs (equal parts of brandy and sweet spirits of nitre); a bottle of diarrhœa or scour mixture (Mr. Leeney advises 1 oz. of trisnitrate of bismuth, $\frac{1}{2}$ oz. of powdered catechu, 1 oz. of powdered chalk, 1 oz. of laudanum, and sufficient peppermint water to make 20 fluid ounces; give one teaspoonful to very young lambs, adding another for each fortnight); a bottle of laudanum (to be used carefully and sparingly); a bottle of castor oil (in cases of constipation); a good knife (curved inwards at the point); Glover's needles and thin tape (for use in cases of eversion), and such other appliances as the shepherd understands the use of (which are generally very few); some vinegar with blue vitriol in solution (to dress the feet of lambs or ewes which have become raw through being on wet litter, etc.);

and a bottle of foot-rot mixture (to dress the ewes suffering from foot-rot, which should be attended to as early as possible).

Lambing, or Yeaning.—When the time of lambing approaches, the sheep must be closely watched, as it may be necessary to give the ewes assistance. Many of the most complicated cases of lambing occur among the first few, as those ewes which lamb prematurely, carry dead lambs, or have other complications, are likely to come on early. The normal cases come at normal seasons. If the ewe comes on naturally, and the lamb comes right, it is generally best to leave all to the course of nature. If the ewe has a small opening, aid may be necessary, and this occasionally occurs in the case of young ewes ; but many troubles arise through too early interference.

Signs of approaching parturition are generally noticeable some little time before the lamb appears. For a few days the milk forms in fair quantity, and the udder is somewhat distended and red. Later the tail appears to rise up. This is a delusion, as its prominence is due to the pelvic bones having parted to allow room for the lamb to pass. Just before lambing the water-bladder appears. If the lamb does not come forward within an hour or two, there is reason to think that something is amiss, though active steps need not be taken for some time.

In the ordinary course the lamb will appear with its nose resting on its two fore feet, the hind legs drawn up under the body ; then little danger need be expected. If help is required, the legs should be drawn out singly ; then stretching the opening with one hand to give the head a better chance of coming out, the body should be drawn forward by the legs, which should be pulled downwards towards the hocks of the ewe. The force should be applied at the same time as the ewe heaves. The lamb is covered with a thin skin (the placenta), forming a caul over the head, which should be removed from the nostrils to allow it to breathe. If the lamb is strong it shakes and sneezes this out, but weak lambs occasionally have not sufficient strength to do so.

Malpresentations.—Sheep which have been subjected to high treatment and close folding are liable to greater difficulties at lambing than are those less highly developed and allowed to roam with little restriction. This is mostly shown by what are known as malpresentations, in which the lamb does not come forward in the normal manner. Malpresentations take a variety of forms, a few of the commonest of which may be explained. The simplest form of malpresentation is where one or both fore legs are turned back, although the body is otherwise in a proper position. In this case the lamb must be pushed back until room can be found to bring the legs forward naturally. This should be done gently, and the advantage of a small hand is realized on these occasions.

" There is plenty of room inside " is a good maxim for the shepherd to bear in mind in all cases of malpresentation.   He should seek for the leg, get his fingers behind the knee, and gradually draw it forward ; and if the lamb is lively it may be well to slip a noose over its leg when brought forward, and then seek for the other leg. Having brought both forward, the lamb should be drawn out without delay, except to give the ewe a short rest if overcome by exertion.   The force should not be too great.   Some other irregularity, such as the head turned back, may not be noticed.   Once, however, the legs and head are in line the work of delivery may be proceeded with.

Sometimes the feet come all right, but the head may be turned back towards the shoulder ; then it is necessary partly to replace the lamb so that room may be found to turn the head into proper position.   The slippery nature of the mucus covering the lamb makes it difficult to grip, and the head must be coaxed round by the fingers until a better hold can be got in the mouth or other convenient spot.   Occasionally a leg may have got over the head, causing an " arm-over-head " presentation.   If the lamb is small and the ewe roomy, gentle movement of the legs may give relief ; but generally it is necessary to shove the lamb back to gain the advantage of more room.   A simple case of one leg back often gives trouble, and in an ewe with little room, space will have to be found by shoving the lamb back far enough to permit the hand to get behind the shoulder.   The legs may be turned under from the knees ; if so, they must be put into proper position.   In fact, the legs are the main cause of difficulties, and the experience necessary to recognise when the legs are in place is of the greatest importance to the shepherd.   To discern between the fore legs and the hind legs is a highly important matter, and to be sure of this and of their relative position is one of the first requisites ; for this reason the shepherd should, in all cases when he finds it necessary to insert his hand, try to impress on his mind the " feel " of the several parts of the lamb.   This is particularly necessary where the lamb makes a rear or breech presentation with all the legs forward (for the back legs are usually those most forward) and the head turned under.   If the fore legs can be shoved inwards the lamb may come away easily, rear foremost ; but it may be necessary to completely turn the lamb to produce a normal presentation.   A rear presentation with the legs backward (practically standing) may be very awkward.   Unaided delivery is almost an impossibility, and the lamb must go forward and be delivered either as a back presentation by getting the hind legs up, or by turning and making a normal presentation.   In putting back a lamb, it is always advisable for the ewe's hind quarters to be higher than the fore quarters. Twins in the majority of cases come fairly well, and are delivered with reasonable care, because they are not of abnormal size. This is not, however, always the case, as sometimes one lamb is unusually large and the other very small. When the lambs are both making normal presentations there is no particular trouble, unless both come forward at once, when one must be shoved back and the other taken out, leaving room for the second. They sometimes lie " head to feet," and one may be taken as a normal and the other as a rear presentation if there is room, or each must be treated as an individual case, according to its position. Care must always be taken not to mix the legs of the two lambs when help has to be given, and many cases where there are twins are erroneously treated because one lamb only is assumed to be present.

When the ewe's time has come, and the water-bladder has appeared, yet no other signs are apparent, as the ewe will not strain, or straining cannot bring the lamb forward, there is often a dead lamb within her beyond the reach of the shepherd. Artificial pains should be induced by means of ewe drinks (sold specially

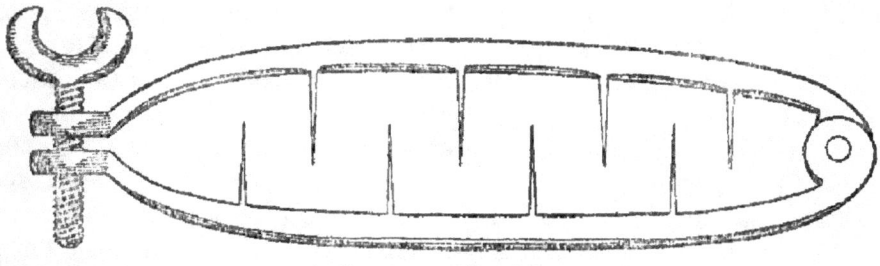

A CLAMP FOR EWES.

prepared for the purpose) in which there is generally ergot. These are not always effective, and the ewe frequently succumbs. Before lambing, ewes are liable to eversion of the uterus, when the breeding bag is forced out. A simple and useful clamp for retaining the bag is to be obtained from Mr. Huish, 8, Fisher Street, W.C. When the bag is protruded it should be put back and retained by the clamp, or by a piece of tape stitched across the shape. The tape should be wiped with an antiseptic (such as carbolic acid and olive oil, in the proportion of one to seven, or one of the advertised carbolised oils) to prevent contagion of any kind. The tape must be cut or the clamp be removed when lambing is imminent, otherwise the ewe will be badly torn, or she may not be able to get rid of her lamb, and die from exhaustion.

As soon as possible after being born the lamb should be induced to suckle. Healthy lambs give little trouble, as they soon find the teat. In the case of a weakly one, the teat should be placed in its mouth and a small quantity of milk be milked into it. It will soon gain strength. It is advisable in all cases to draw the teats to insure a clear passage for the milk, as sometimes they are blocked with dirt. Whilst handling ewes, the shepherd should keep his hands well washed, or, going from a foul one, he may take disease to a healthy one. This is especially necessary in cases where there is a dead lamb. If the sheep is long in lambing, or gets the skin or womb ruptured, or has produced a dead lamb, a good quantity of a mixture of carbolic acid and olive oil in proportion of 1 to 7 should be injected, and the shepherd should rinse his hands in the mixture. Dead lambs should be buried, as also should the afterbirth of any lamb which is born prematurely. Dogs should never be allowed to eat dead lambs before they are skinned, as they are liable to acquire a taste to satisfy which they may become sheep worriers.

Prevention of Infectious and Contagious Diseases in the Lambing Pen.—Mr. Harold Leeney, in a very instructive paper on " The Lambing Pen," which appeared in the *Royal Agricultural Society's Journal* of March, 1897, says in respect to infection in the lambing pen :—" The bulk of losses at lambing time occur through want of systematic disinfection of hands, implements, appliances, and buildings. Nor is it any disproof of the efficacy of antiseptics to point to a record of success. The infectious elements are not always equally active, and we have to consider whether those ' unlucky ' or bad lambing seasons might not turn out a great deal better if attention were given to some of the simple precautions about to be mentioned. The remedies and medicaments with which the shepherd should be provided are not many, but their constant use is of importance. They include a bushel or two of lime, fresh burned and ready for slaking, when it is desired to disinfect the earth ; a gallon of Jeyes' fluid, or other similar preparation that will readily emulsify in water ; a quart or two of carbolised oil, in proportion of 1½ oz. of carbolic acid (pure) to each quart of olive oil ; a cake or two of carbolic soap—preferably Calvert's or some maker's whose guarantee that it shall contain 15 per cent. of acid can be relied on—the common soaps are variable and cannot be trusted ; half a dozen penny sponges ; some clean rags, such as old calico underclothing ; and a bowl or metal pail (not a wooden one) for washing hands, &c. These are the essentials for disinfection and for antiseptic purposes."

In relation to this paper of Mr. Leeney's it can with confidence be stated that it was an epoch-making one in shepherding ; for it instituted a general attention to sanitary and hygienic matters which were practically unknown to shepherds and flock-masters previously. It has resulted in decrease in the losses of ewes and lambs which it would be difficult to compute, but it is very great.

Daily Work in the Lambing Fold.—The general procedure in the lambing pen can be best explained by describing the shepherd's daily work in the pen. In the case of large flocks, the shepherd should not be given other work to do during the busiest season of lambing. He has to give attention to the ewes both night and day, and get his sleep as best he can. If he does not get reasonable rest some part of the work is neglected, and a small amount of neglect may result in large losses. We have seen several instances of this. Beyond an occasional look at other portions of the flock, he should not be troubled with them, and if he is hard driven he should have plenty of help with the ewes. The loss of an ewe involves as great a loss as would pay a labourer's wages for three or four weeks, and many lambs worth several shillings each can be saved every year by proper attention.

The first thing for the shepherd to do in the morning is to attend to the lambs which have been born during the night, to see that they suckle. Then the ewes about to lamb should be looked after. It is more convenient to put these in separate small pens around the yard, as they can then be found and handled without loss of time. Older lambs should then be looked to, and those which are weakly should be suckled or fed from a bottle. The ewes should then be fed with roots, and the backward ewes be taken out to their fold. Having had his own breakfast, if his attention is not required with an ewe about to lamb, he should litter the pens afresh. Foul pens are very productive of foot-rot and other far more deadly diseases. The ewes require their corn and fresh hay; lambs require attention, particularly those which are weak or where the supply of milk is short. In case a mother dies, or cannot support her lamb, the lamb must be fed from a bottle or from ewes which have a surplus supply. Where an ewe possessing plenty of milk loses her lamb, she should be given another. When the ewe and lamb are placed in a small pen, the difficulty of making her take to a fresh lamb can be overcome in a few days if patience is exercised. She must be tied up, otherwise at first the lamb will get no milk, and it is often necessary to stand beside her for a time to let the lamb suckle. The best means of restraining her is to drive two stakes firmly into the ground, and then to place her head between them, allowing room for her neck to move up and down, but keeping her head firmly secured between them. Room should be given on either side to allow the lamb to escape in case she tries to trample on it. If she is exceedingly savage it may be necessary to place another stake on either side of her, so that she cannot knock about. If the skin of her own lamb is placed on the foster lamb, or if the latter is rubbed with it, she will recognise the smell and will be more likely to take to it. It is not an uncommon sight to see lambs very much fouled behind through putting the skin on whole and leaving no escape for droppings. This filthy practice should not be allowed.

As soon as the ewe recovers from lambing she should be dressed if she has foot-rot, to prevent it from spreading, and the lamb's feet should be kept clean and healthy. At the least sign of lameness the lamb's feet should be cleaned and wiped dry, then a small quantity of a mild caustic (sufficient to moisten the abraded parts) should be placed between the claws to harden them and prevent foot-rot from being established. A mild caustic of 3 oz. of blue vitriol (sulphate of copper), dissolved in a pint of vinegar, effects this without inconveniencing the lamb, and is preferable to stronger caustics. The sheep require feeding as night approaches, and should receive roots, hay, and the second portion of the corn. The backward ewes require bringing in for the night, and those which have already lambed sufficiently long to go out into the fields during the day should be brought in also. The shepherd has to prepare for a long night, as early in the season the days are short. The last thing to be done in the evening is to make sure that all the lambs are well fed, and special attention should be given to those which require suckling from their mothers or feeding from a strong-necked bottle. A short piece of elder, from which the pith has been extracted, inserted in the cork of the bottle, makes a convenient and safe mouthpiece; but it is far better to use rubber artificial teats, which cost little and are easily cleaned, though an occasional wash in an antiseptic should be given.

Having got his sheep safely folded, the shepherd's attention is chiefly required among the ewes which are likely to lamb, so that he may be at hand to give assistance where necessary. He must get his sleep as best he can, and for this reason the advantage of living near to the yard, or better still, in a temporary hut, is evident. It may be necessary for him to be about nearly the whole night when lambing is going on rapidly. At other times it is sufficient if he looks round every two or three hours. The advantage of having a large number of small pens handy to turn in the ewes as they look like lambing is especially marked at night, when the only assistance the shepherd has in finding them is a lantern. It is advisable to mark each single lamb with a spot of ochre, and each twin with two dots, so that they may be recognised. Confusion is often saved by doing this. If ewes get low in condition, and weak through troubles at the time of lambing, they should be given easily digested and nutritive food. Concentrated strengthening powders, specially prepared for the purpose, are valuable aids to the shepherd in such cases, and should be kept in readiness. Strong, warm gruel of oatmeal, ginger and strong ale is serviceable when these are not at hand. The gruel may be made of a pint of ale, 1 oz. of oatmeal and 1 oz. of powdered ginger.

CHAPTER XVI.

AFTER-MANAGEMENT OF EWES AND LAMBS.

In the Field.—It is a most important matter to get the ewes off the foul litter of the lambing-pen as early as possible ; therefore as soon as the lambs are strong enough, they should be taken out of the fold during the daytime, and be put on dry land. The lambing-fold, as has been stated, is very likely to produce foot-rot

SHELTERING CLOTHS IN USE.

and joint-ill, and these should be avoided by every possible means. Shelters, formed by hurdles stuffed with straw, should be set about the fields to protect the lambs from cold winds and wet, and should be arranged so that the lambs can take refuge from the wind whichever way it blows ; or sheltering cloths may be used for the same purpose.

The ewe's food should be liberally supplied, so that she may provide a good flow of milk. Among the foods which particularly conduce to the flow of milk are oats, peas, decorticated cotton cake, linseed cake, dried grains, with ensilage, grass and roots. Wheat and barley are also valuable, although some hold that the latter is not. It is best to give them in mixture, rather than singly, as the admixture acts beneficially. The lambs cannot digest food other than milk for a few weeks, during which, of course, they are dependent on their mothers. Maize, barley and turnips do not make a good mixture, as the nitrogenous constituents are lacking. Oats are especially valuable, but for some little time before weaning the quantity should be restricted, as from the extent to which they excite the flow it is difficult to dry the ewes as rapidly as is safe. All corn should be sweet and free from mustiness, as sour corn or cake makes the milk unwholesome. From $\frac{1}{2}$ lb. to 1 lb. of a mixture of corn and cake should be given daily to the ewe, according to the supply of green food and the purpose to which it is desired to put the lambs, though, especially with big ewes like the Hampshire, where it is wished to do the lambs at high pressure, as much as 2 lb. of cake and corn are given. If they are to be fattened off quickly as fat lambs, the full quantity should be given whilst they " take their corn through their mothers "; afterwards it can be gradually withdrawn from the ewes and given direct to the lambs. If the lambs are to be kept on to be fattened off as tegs, the smaller quantity of corn will suffice. The lambs, however, always show a profitable return when the " lamb's flesh " is kept on them as long as possible.

The lambs should be induced to feed as early as possible; if a small quantity of finely ground linseed cake or pea husk is placed in a small trough outside the ewes' pen, and if a properly constructed lamb hurdle, such as will allow the lambs to pass through, but keep back the ewes, is placed near, they will soon find their way to it, and learn to feed. The lambs should always have an opportunity of feeding in front of the ewes, whether they are on grass or roots. Nothing is more suitable for lambs than the young tops of turnips, kohl-rabi, or kale and rape. The former are valuable early in the season, but get too old when they run to flower; rape follows, and kale is extremely useful throughout the spring, but more especially from March onwards, when the plants have thrown their spring sprouts. Failing young sprouts or fresh grass, mangels cut into fine slices are a good substitute. It is important that the lamb feed shall be fresh, and even in the case of corn or roots fed in troughs only small quantities should be given, so that the lambs may clear them out quickly. They will not return to stale food, and if any is left over it should be cleared out of the trough and given to the ewes, which are not so particular. We have found nothing which keeps lambs in such healthy condition as the husk of peas—"pea chaff" as it is often called. Though not containing so much nutriment from an analytical point of view as some other foods, it appears to be easily digested, and the digestive organs are kept in vigorous condition by it.

It is astonishing what a large quantity of rich corn and cake lambs are capable of digesting when they receive their mother's milk. The one seems to aid the digestion of the other, and they do not suffer from overdoing, as is frequently the case in after life. With the improvement in systems of feeding, and the greater aptitude to fatten, sheep of the Down breeds particularly mature early.

On Grass.—Less attention in the matter of feeding the ewe is generally practised as the lamb becomes more independent of her. It is a mistake to let the ewe get too low in condition ; at the same time, extravagant feeding at this period is not warranted. In the management of a breeding flock economy, consistent with the health of the sheep, must be exercised, or the cost of the lamb becomes excessive. An ewe when drafted lean from the flock after breeding three or four lambs is worth no more—often less— than she was when put in the flock as a theave. She spends each year in producing a lamb, and the lamb's cost at the time of birth is that at which the ewe has been kept through the year, with slight additions for percentage of loss and a proportionate share in the cost of the ram. On grass this expense is slight, and on arable land it is dependent on the cost of raising crops.

A lamb (expecting that by good management there will be a good percentage of twins) ought to be born costing not more than ten shillings, although it may pay in highly bred show flocks for it to cost twice that sum, a large portion of it going in the cost of the ram, though on hill land the lambs of small breeds must cost considerably less than 10s. It is fair to charge the lamb with the cost of the corn the ewe receives for two months after lambing, as it is taken that the lamb is receiving corn through its mother. When the lamb gets its own living, however, the ewe ought to be costing little more than at other seasons. The food of the ewe should not cost more than twopence a week throughout the year, as other incidental charges bring up the total to ten shillings. The advantage of getting a good percentage of twin lambs is easily apparent, as is the necessity of keeping as many as possible alive. It does not pay to let ewes get feeble from lowness of condition, as some permanent injury is often brought about by so doing. Except, however, for a short period after lambing, the ewe must be regarded as the farm scavenger, often utilising that which it would be unprofitable to keep other sheep on. The first picking of the food should therefore be given to the lambs from the time they are strong enough to feed.

Tailing and Castration.—When lambs are about twelve days old the ram lambs not intended for breeding purposes should be castrated, as at that period they suffer very little from it. In some districts the operation is left until nearly weaning time, when emasculation, followed by searing, is adopted. The simplest method is to cut a slit in the side of the scrotum and press out the testicle, then to draw it out with pliers or by the teeth, castrating tool, or by the thumb and finger, or to cut off the bottom of the scrotum as may be preferred. In the long wool breeds it is often desired to diminish the size of the scrotum, and then cutting off is advisable; most Down breeders, however, like to have the full purse left, as when well filled with fat it is regarded as a good indication of condition. My own practice always is to draw by thumb and finger, and if the knuckle of the hand is pressed hard into the body of the animal it gets steady leverage which ensures full control of the drawing In doing this over many years I cannot call to mind a single loss. Old lambs, that is when three or four months old, are better seared.

The operation of tailing the lambs should be done at the same early age, when both sexes are dealt with, and if the skin is pressed back towards the body, the operation causes little pain, and the wound soon heals. A dab with a sponge holding a mild antiseptic solution is a safeguard against septic poisoning.

Castrating and tailing are best done on the evening of a fine day, and the sheep should be kept quiet during the night. With older sheep the wound is generally seared, but on young lambs a little tar to keep away flies is all that is necessary. Tailing is done to promote the comfort of the sheep subsequently. Breeds with much wool on the tail are liable to get very dirty tails when the animals, through change of food or other causes, scour. A long tail under such circumstances becomes a source of discomfort and ill-health. The foul matter collected about it is a great attraction to the flies which produce sheep maggots, and it is almost impossible to keep them away. Breeds which live almost exclusively on thin mountain pasture and produce wool of a loose and hairy nature are less liable to injury in this way, and their tails are frequently left uncut. The tail is the natural covering and protection of the hind-quarters, therefore on exposed hills it is advantageous for the tails to be left in their natural state. Lambs intended to be sold as fat lambs must never lose their *lamb flesh.* By lamb flesh is meant that plump sleekness associated with the appearance of a young healthy lamb. This rapidly disappears when, through ill-health or shortness of food, the lamb becomes poor and pot-bellied, and once lost can never be wholly regained. It, however, may be kept up until the lamb is some three months old if proper feeding and care are bestowed on it. To some extent it may be continued longer, but unless fed at high pressure the frame attains the looser appearance associated with young sheep. Although uncastrated lambs grow faster for a time than do those that have been castrated, yet castrated lambs make the best fat lamb on the same amount of food ; they handle better, and die better.

Older Lambs.—A variety of green food is beneficial to lambs ; in fact they can hardly receive too varied a diet. The finest lamb feed of any individual crop is white clover, but care has to be exercised in its feeding, as in place of ordinary digestion it is liable to produce a large quantity of gas in the stomach, causing what is known as *hoven* or *blown.* At times the stomach is so distended that it bursts, and the lamb quickly dies. The first sign of hoven is usually frothing at the mouth. If this is detected no time should

be lost in getting the lambs out of the field ; the exercise of walking is beneficial, as it helps to move the gas. Should a sheep become so bad that fears of its safety are entertained, it should be relieved by the insertion of a trochar and canula, when, on the withdrawal of the trochar, the gas will speedily escape. The canula can then be removed, and the wound will

INSERTION OF TROCHAR AND CANULA.

speedily heal. Special care must be exercised for a few days to prevent recurrence. Failing the proper instruments, relief may be obtained by means of a stab by a knife. The hole made may be kept open as long as desired by inserting a stout quill. The exact position to insert the instrument is equi-distant from the hip-bone, the last rib, and the lateral processes of the backbone, and the direction should be nearly horizontal with the point directed slightly downwards, otherwise the kidneys may be injured ; nor should it be done on the right side, as then other organs are endangered. The point at which to thrust the instrument is shown in the centre of the triangle in illustration, and the direction of the thrust is also indicated. Clover being very succulent, is the most likely green food to produce hoven, especially when hoar-frost or moisture are on it. The best means to prevent this is to put the sheep on the clover when their stomachs are well filled, to prevent their eating too ravenously. Other foods occasionally cause it, and it is not restricted to lambs, as sheep of all ages may be affected with it, especially in windy weather, though why windy weather should have any influence is not apparent. The shepherd, however, should always be on his guard.

As the autumn-sown catch crops come in, the lambs require them, and if they can be changed two or three times a day, so as to get variety, they thrive better. Grass, or seeds, green tops of rape or kale, and vetches or other autumn-sown catch crop, make an excellent diet. When cabbages come in they are very valuable, as there is scarcely anything on which lambs thrive so well. The great aim is to keep the lambs from diarrhœa or scour. They lose flesh rapidly when they scour, and it takes a long time to get them into a thrifty condition again. Sweet food is absolutely necessary and nothing tends so much to rectify the bowels as a small quantity of winter barley, grown as a catch crop. It would pay every sheep farmer to grow a small piece of this crop every year, as it gives a good return, in addition to its medicinal properties. Another remedy for scour is to allow the lambs to nibble the shoots of a whitethorn hedge. The astringent properties of the shoots have a decidedly beneficial effect.

When the lambs are weaned they may be regarded as tegs. Until recent years it was not until autumn that this title was given them, but owing to their more rapid maturity under present systems of breeding and management, the career of a sheep is much shorter than it was. Except among the less improved breeds, existing under conditions which permit little alteration in feeding, as on hill land, three or four-year-old wethers have become a thing of the past. In the more highly developed breeds the sheep pass through the comparative stages of lamb, teg, and wether, in as little as ten months. The lambs are sometimes shorn in July, and so lose their teg locks; the short wool which grows in the autumn gives them the appearance of wethers, and they go to the market as wethers.

CHAPTER XVII.

TEGS.

Tegs in Summer.—Regarding the lambs from weaning time as tegs, and speaking of them as such until they are a year to a year and a-quarter old, and in the natural course are shorn, the object is to get them fit for the butcher from November until spring. On grass-land farms they spend the greater portion of the time at grass, and it depends on the quality of the grass and the amount of corn they receive throughout their career, and on the hay and roots brought to them in winter, as to whether they go out as tegs, or are well on to be fattened out as wethers in the following summer. On farms where there is a large quantity of grass, but a fair proportion of arable where roots can be eaten off on the land in winter, they get a run on the stubbles in autumn, and probably white turnips, rape, or cabbages as a night fold in addition to pasture. In winter and early spring, until fresh keep appears, they are folded on roots, receiving hay and corn if not intended to be fattened out as spring tegs; and corn is less often given if they are only required to grow through winter and be fattened out when keep becomes more plentiful in summer. Good grass land should carry from ten to twelve sheep per acre from the time it is fit to stock in spring until autumn, some being drafted out from time to time as they fatten. On moderate pasture it is commonly reckoned that an acre will carry two ewes and their offspring, which may be taken as five sheep in all.

Summer Keep or Feed.—The lambs, on weaning, should be dipped in one of the well-known dipping solutions to free them from vermin which may be on them, and to prevent others from locating themselves. This also has the effect of rendering them less likely to be struck by the fly which lays its eggs on the wool, the eggs speedily developing into maggots. The feet must be kept sound by paring, and if disease breaks out they must be dressed with a caustic foot-rot solution, of which there are many. On farms where arable land greatly preponderates, the tegs are kept to a great extent on the arable land at all times, as the grass is required for other purposes; though, if other kinds of keep are stale or short, a fresh run on the meadows is desirable. During the early part of summer, when they are first weaned, it is generally easy to find fresh, sweet food for them, but as the season advances it becomes more difficult. Unless lambs get sweet keep they soon "go wrong." "Going wrong" conveys a special meaning to sheep-keepers, as it implies that there is a derangement of the digestive organs, which is shown by the lambs becoming constipated or too open, more usually the latter ; the wool becomes dry and harsh, and instead of lying smoothly and sleekly, is rough and broken, while they lose flesh, and no amount of food appears to do them good. This state of affairs is very commonly brought about by stale or sour keep. Food may be freshly grown, succulent, and to all appearances favourable for animal food, and yet be distinctly injurious to young sheep. "Stale food" is most commonly found growing on land which a short time previously was fed off by sheep. It becomes soured by the droppings of the sheep, though in what way is not definitely known, but it has been suggested that the souring is due to a larger quantity of magnesia being taken up by the plants from the manure recently deposited. More probably, however, it is caused by parasitic worms or specific germs. Rape fed off by sheep in summer is soured ; yet after standing over the winter, and being subjected to frosting, it becomes sweet and wholesome, no magnesia having been dispersed from it, though it may have undergone some chemical change which has not been discovered. Temporary leys in which there is a large proportion of clover are dangerous in this respect. However, if the young sheep are not turned on again until after a crop has been taken off by mowing, no harm will as a rule come of it. It appears that the first flush of growth after feeding is the dangerous portion. Grass land is influenced to a lesser, though still a noticeable, degree. Cabbages fed off by sheep and allowed to sprout again are dangerous to young sheep if fed before winter. Old sheep, having more robust systems, do not feel the effect in a similar manner. In feeding off his crops during the early part of the year the farmer has to look forward and arrange that there will be a supply of sweet fodder throughout summer. The aftermath of seeds and meadows are a source of reliance, and cabbages and kale are of great value during the period between hay-time and the end of harvest, when there will be fresh stubbles, and probably white turnips, available also. Early rape may be fit to stock, though only that sown very early will be sufficiently grown. Autumn-sown vetches should be available until July, when spring-sown vetches should come in. If a good succession of crops like these is grown there should be little serious difficulty in finding ample sweet food for the tegs until autumn. Our own experience has shown us that there is nothing equal to cabbages and kale as food from July to September ; they are less affected by drought than are most crops, and, as it is not of so much importance that they are fed at a definite period, they may be held over for a time, or some other portion of the cropping may be diverted to other purposes. Whatever provision is made, however, it is imperative that the young sheep shall not be forced to go on to stale keep. That various kinds of worms have a direct influence on unsound or stale keep, however, appears to be pretty certain, and attention is directed to the discussion of the subject in the Veterinary Section.

Concentrated Food.—Through summer the sheep receive cake or corn in accordance with the views the farmer holds as to the time at which they shall go to the butcher. Those which are to run the whole year round and be sold as wethers require none, though we like to give about half a pound of pea-husk per day as a corrective. Ewe tegs to go into the flock can be treated similarly. A quarter of a pound of corn or cake per day is sufficient where the tegs are not fattened out until spring. Where they are being fed fast to be fattened out from November to January, from half to three-quarters of a pound may be necessary.

Sheep do much better when they receive a fresh fold daily, and they waste less food. A good fall-back, however, may be allowed, though it is generally better for the ewes to clear up behind them than that the young sheep should have to graze too closely. This, however, is to an extent dependent on the object in view.

Tegs in Autumn.—In autumn the food varies; the stubbles and young seeds afford a fresh run. Some farmers object to allowing sheep on young seeds, but, except on thin, weak plants, or on heavy land in wet weather, when the sheep make deep footprints which are left as cups to hold up water during winter, the treading the land receives is beneficial to the young plants, as the earth is more firmly pressed round the roots, and there is no need to graze so hard that the heart of the grasses or clovers is gnawed out. The ordinary practice of rolling does not do nearly so much good, because the soil is not so regularly tightened about the roots. The Drumhead cabbage, white turnips, and other crops come into season now.

The winter is the season for feeding hard roots, such as turnips, swedes, and kohl-rabi. It is important, as far as possible, to get the sheep gradually broken to turnips, and not to take them straight from grass, seeds, or other foods, and put them on roots. This sudden change of diet often causes great loss of life, while many sheep receive a check in growth and fattening from which they do not speedily recover. So well is this recognised that in buying-in in autumn, those sheep known to be broken-in to roots are worth a shilling to two shillings a head more than are

K

sheep in every way similar except that they have received no roots, but have come straight from grass. Cabbages and rape are good stepping-stones to roots, and white turnips (probably owing to their ripeness) and rape are less injurious than swedes. It is advisable to give very small quantities of swedes at first, and to give with them a plentiful supply of dry food, such as hay or hay chaff. The importance of dry food throughout winter cannot be too vividly remembered ; nor should the season be allowed to advance too far before it is given. In the case of either fattening or breeding sheep, a plentiful supply of dry food is the most effective means of preventing losses in winter.

Tegs in Winter.—When the tegs get on to roots the turnip-cutter should be set to work. Cabbages, kale, rape and white turnips are soft enough to gnaw, but swedes and kohl-rabi should be sliced. In the case of sheep which it is desired to get out very quickly, it is preferable to slice white turnips, as they satisfy their appetites more easily, and have thus a longer time to rest and digest their food. The saving in food repays the cost of labour, and the sheep thrive better.

The corn should be increased as the sheep become fatter and it is desired to get them quickly to the butcher. Under ordinary circumstances a pound weight per day is all that it is advisable to give. Large quantities are sometimes given, but the risk of loss from overdoing is great. " There is another sheep dead this morning, sir," is the unwelcome news every farmer who fattens large quantities of sheep at high pressure knows only too well. It occasionally results from eating too many unripe roots, but generally from what is known as " making too much blood." It is, indeed, a form of paralysis caused by too much nitrogenous matter in the blood, this being brought about by a diet containing an excess of nitrogen. This presses on the brain, and the sheep rapidly succumbs if assistance is not given. The obvious remedy is to weaken the blood, and the shepherd does this by drawing off a quantity as soon as possible. The corn must be withheld, and only gradually returned to the sheep, as when one sheep shows signs of excess it is probable that others of the flock are in danger. It indicates that the corn is too nitrogenous, and that this must be changed for a mixture which is more starchy. A sheep with paralysis appears listless, and lies in a helpless condition. As soon as signs of this are apparent, no time should be lost in relieving it of blood. Half a pint may be taken away with safety. The best vein to open is that on the side of the face, a little below the junction of the eye orbit and nose. Some bleed by cutting the ear ; others from the vein on the inside of the leg, a little above the knee. If left too long the sheep will soon die. It is held as being unscientific to let blood ; but I have saved too many sheep's lives by this prompt action to give up the practice. Aperients and medicines which contract the size of blood corpuscles are very valuable, but they are not always available, nor are they as prompt in their action, and it is often a case of immediate treatment or death. However, a careful shepherd never allows a sheep to die a natural death ; when he sees that there is little chance for it, he kills it and dresses it, so as to make it marketable. It is now illegal to sell the flesh of sheep " killed to save their lives," but anything more absurd than to prohibit in the case where an animal, otherwise in good health, is killed because of a quickly developed blood pressure on the brain, and before the system is affected, it would be difficult to imagine. It is different where the sheep had been drugged. The advantage of a well-balanced mixture of corn is easily recognised when the danger from one in which too much nitrogen is present is understood.

All roots should be sliced for fattening sheep when once they get on to the harder kinds. It is best to eat the swedes in mid-winter, and to follow these with kohl-rabi after the turn of the year ; though the early Small Top Kohl-rabi are essentially autumn food. No farmer should allow the first week of December to pass without having a portion of the earliest and ripest swedes—sufficient to supply the sheep for a month—safely clamped. Neglect to do this is the source of great waste during frost ; the roots are more difficult and more expensive to get up when frozen in the ground, and are less valuable as food owing to their excessively cold condition ; while in very severe frosts they are totally des-troyed, and great difficulty is experienced in getting a supply of food during the remainder of the winter season. This advice may seem unnecessary as it is now over twenty years since there have been December and January frosts which have imperilled the swede crop. Yet in thirteen years previously there were seven seasons which were so severe at that time that the losses by those who had not clamped their swedes were very seriously heavy. Experience in the long past has shown that there are periods lasting over several years of mildness or severity, and cold winters will doubtless return. As a rule turnips may be pulled by hand. If very tightly in the ground a turnip-pecker is necessary. This is also needed where sheep gnaw the roots, as, if the cups are not pecked up, they are wasted. The hardy variety of kohl-rabi, particularly if not overripe, will withstand all but the most severe frosts, and need not be clamped. If ripe, the plants may be got up and clamped in the same way as swedes. If the tegs have been liberally treated throughout the season they should be fattened off and sold from time to time as they become fit for the butcher, so that they are sold out by the end of March. Those which have to be run on and sold as wethers should give little cause for anxiety ;

they are not fed at high pressure, but receive sufficient food to keep them in a healthy growing condition, when little risk is run. The distinction between the fattening-out of tegs and wethers lies chiefly in this : tegs receive corn practically throughout their career, and wethers only at the end. Teg-fattening involves considerably outlay in the purchase of feeding stuffs ; wethers are fed almost entirely on the produce of the farm. A portion of the tegs are often kept for selling out between March and June ; that is, the last of them are finished off on the first growth of grass or seeds. These, as a rule, are not fed fast during the autumn, and probably receive no corn until Christmas, when they are given a small quantity ; and this is increased slightly from time to time as they appear to need it, a much larger quantity being given for the last month before they are turned out fat. These tegs are often shorn before being sent to market.

Tegs kept through winter on grass are generally maintained at little cost. There are instances where it would pay to give more food, particularly in those cases where sheep are sent out to grass on agistment to distant farms. The fear that the corn supplied or paid for may not be given to the sheep probably restrains many from providing it ; but, considering the loss of life, and the fact that many of the sheep weigh less after the winter grazing than when they are sent to it, something more is obviously required. Where sent out to winter on roots, some dry food, such as hay, ought to be provided, otherwise the progress is very slight in relation to the money spent on their maintenance. The small additional expense makes a large difference in the growth and condition of the sheep. It is a foolish policy to spend ten shillings on wintering sheep in such a manner that they do not improve during that time, when half a truss of hay, at 1s. 6d. per truss, will supply a quarter pound of hay per day for nearly four months, and make a marked difference in their condition : half a pound per day is, of course, better.

Pasturage is the natural food for sheep, and all sheep do well on sound grass ; though in winter the breeds which have been highly developed so that they will fatten out within a year from birth require better food than even good pastures will provide in winter, unless much additional food is given them. But strictly grass breeds will maintain themselves in a store condition if the pasturage be plentiful. The carrying powers of nearly all pastures can be greatly increased by the help of suitable manuring. Poor land pastures, especially those on thin light land, or heavy land exhausted by hay crops or by carrying dairy cows for a long period, nearly always require phosphatic manures, and frequently potash or lime. With an absence of these the leguminous growths so beneficial to sheep are sure to be short, and the grasses themselves will be of inferior type and growth. Sheep-breeders should more often recognise that their pastures would produce far more meat, bone, and milk were suitable manurial assistance given. Generally, the acid-treated phosphates are most suitable to light land, and untreated, such as basic slag, on heavy or wet ones. Dry pastures are always healthier than wet ones, consequently drainage should be good. Sheep should be run thinly over pastures, and be frequently changed from field to field.

First Feeding of Leys.—Good grass pastures and temporary leys of grass, clover, sainfoin, and lucerne, afford good food for wethers, and they can be kept very economically thereon. While food is plentiful, particularly on the first feed-over, the sheep thrive with very little corn. When I was managing the Woburn experimental farm for the R.A.S.E., I had a striking illustration of the value of the first feeding of seeds as compared with subsequent feedings of the same crops. For the sake of experiment the crop was divided into plots an acre in extent, and the sheep received cake in one instance, maize in another, and on two acres no extra food. During the five years over which the experiment was tried in different places, those which received no corn increased as much as those which had it; but in the second feeding-off each year, although the food looked as fresh as during the first feeding, and the crop was as heavy, those which received corn went far ahead, yet to all appearance the clover was as good on one plot as on the others. As far as the feeding went the cake and corn were wasted in the first feeding. Nothing different from common practice was done, so it is highly probable that this has been the case in feeding off much similar cropping in other places. Close attention to the " doing " of the animals would probably show that much corn is wasted on old sheep at this time. Sheep can only make use of a certain amount of nutriment in whatever form that nutriment is supplied; and if the crop supplies all that is necessary it is obviously waste to give them anything additional. When the clover is stale, although luxuriant, as in the second feeding-off, the advantage of corn is at once apparent. Those who have a convenient weighbridge might easily put the matter to the test. A skilled sheep-keeper and feeder knows by handling whether or not the sheep are thriving satisfactorily; but every sheep-keeper, no matter how long his experience, is not necessarily skilled, although he may have a good general knowledge.

CHAPTER XVIII.

SHEEP SHEARING.

Shearing.—Wethers supply a large portion of the summer and autumn mutton. Fat lambs are in demand throughout hot weather, but the greater portion of the sheep killed are wethers, largely composed of those which are bred in the later lambing districts, particularly the grass districts. The summer management of these is simple ; they are stronger in constitution than are tegs, and they therefore are less inclined to be affected by sour keep or excessive feeding. The shearing of the wool turns the teg into a wether. In the south-east of England the term teg is retained after shearing ; but this is the only exception. Before shearing it is customary to wash the sheep to relieve the wool of the greater part of the dirt, and to give it a more marketable appearance. Of recent years a number of advocates for selling the wool in an unwashed condition have sprung up, and they have found followers. It is urged that it is to the advantage of the farmer, who, though he gets less per pound, is more than compensated by the greater gross weight paid on. This is opposed to all other experiences. The buyer knows more about wool than a farmer with a limited experience is likely to. The object in offering goods for sale is to make them appear more marketable. Dirty potatoes, unclean wheat, ungroomed horses— in fact, any commodity which is not offered for sale in an attractive form—will not realise so much as when properly prepared for sale. Why, then, should wool ? If a farmer can find an inexperienced buyer who can be misled as to the quantity of dirt in and on the wool, he may gain an advantage, but such buyers are not commonly met with. If there is a doubt as to the actual condition of any article, and if the professional buyer has reason to believe he is running risks in buying, he does not buy confidently, but buys within the actual value, so that he may cover his risk, and thus the seller is the loser. It is to the advantage of the wool-buyer to purchase wool in a crude condition because his special knowledge places him in a better position than the seller with less experience can hope to be. He will not give more than it is worth, but will probably pay less. The question of washing or not washing is mainly controlled by the system of wintering the sheep, and, to some extent, by the openness of the wool. Sheep kept on arable land, especially on sandy land, get much dirt into fleece, and it is almost impossible to shear them properly unless the dirt is washed out. Sheep kept altogether on grass land, however, collect little dirt on their skins, and the need for washing is far less pronounced ; and where this happens, washing is less practised.

Sheep washing is performed in various ways, though most commonly in running streams, the sheep being thrown into it to soak for a few minutes, and then well rubbed with long-handled scrubbers, the operator standing on the bank. Advantage of a bridge over a small brook can be taken to dam back the water, and a landing place cut in the bank a few yards up stream, a washing place being thus very quickly formed. When large numbers of sheep are to be washed it is desirable to give the washing hole a more permanent form. It is necessary that the water be deep enough to allow the sheep to be thoroughly immersed. The sheep should be dropped in rump first, as the water thus breaks through the wool more easily. Particular attention should be paid to the back about the loin, as most dirt accumulates there. The most effective sheep washing place is found where water can be brought from a higher level by means of a trough, as in the case of the overshot water-wheel, as the falling water forces out the dirt more thoroughly, and less scrubbing is required. In such cases it is common for the shepherd to have a tub sunk convenient to the spout, in which he can stand dry, so that he can hold the sheep in the required position. Occasionally sheep are washed in tubs of hot water, but this is generally practised in warmer climates than that of England.

At the same time, it is a mistake to shear sheep after washing until the " yelk " is up again. Here the buyer has a point. The natural yelk, an excretion from the skin, is beneficial in the subsequent treatment of the wool : this is washed out, and it does not rise for a few days after washing ; the time depends on the temperature, it being quicker in warm than in cold weather. The farmer loses the weight of the yelk if he does not allow it time to rise. Washing renders shearing easier, and the sheep is more marketable than when shorn badly, as it must be when its back is full of dirt, for the shears will not keep a good edge, and they will not face the dirt. In some districts, however, shearing is so badly executed that, dirt or no dirt, the sheep are roughly turned out. Sheep shearing is performed by the ordinary shears, or by the mechanically driven shears made on the principle of horse clippers, actuated by any convenient power, such as steam, horse, hand or water ; but more generally by oil engines, except where very small plants are used. There is considerable diversity in the manner of shearing by hand. In some districts the shearer mainly stands whilst working : in others he kneels, the sheep in these two cases resting on the ground ; in others the sheep is laid on a bench, or in a cradle, whilst the shearer stands. In some the legs are tied, in others not. In some districts the shearer cuts longitudinally from end to end of the body : in others he cuts right round, starting at the belly and going over the back down to the belly on the other side : in others the cutting is done against the natural fall or hang of the wool. A very skilled man can, however, make quite good work with only one-hand shearing when the wool is very open, as on some long-

SHEARERS, SHOWING THE THREE POSITIONS.

woolled sheep. The latter makes the best work. The sketch shows three representative attitudes in shearing : (1) opening round the neck ; (2) the middle of the body ; (3) the hind quarters. The sheep should be perfectly dry at shearing, or the wool will not be safe to store. All straw, twigs, &c., should be carefully picked out of the wool, and dirty locks trimmed off. The opening up of the work is performed in slightly differing ways, but the wool about the head should be trimmed off first ; some shearers continue to shear round the neck down to the brisket, merely laying back the wool with the hands : others make a cut up from the brisket to the head to afford an opening. First-class work requires the shearer to be equally dexterous with either hand, so that he may at all times meet the fall or natural hang of the wool.   The difference in shearing with or against the fall of the wool is as marked as that of cutting a laid field of corn with and against the direction in which it lies. A man who can work with but one hand must work one side with the fall instead of against it. The wool hangs vertically from the spine, except on the belly, where it tends towards the opposite direction; thus in shearing the near side, to obtain the best results, the work has to be done with the right hand from the belly to the spine; and on the underline with the left hand from the line of the navel to the outside of the underline. In ordinary work, the belly being out of sight, the wool is generally cut off without regard to this. It is usual to shear all round the neck to the shoulders, after which one side is completed before the other is commenced. A novice finds considerable difficulty in holding the sheep conveniently. The sheep should at all times be held in such a way that the portion being clipped stands out prominently, as then the skin is taut; whereas when loose it lies in wrinkles, which are likely to be cut.

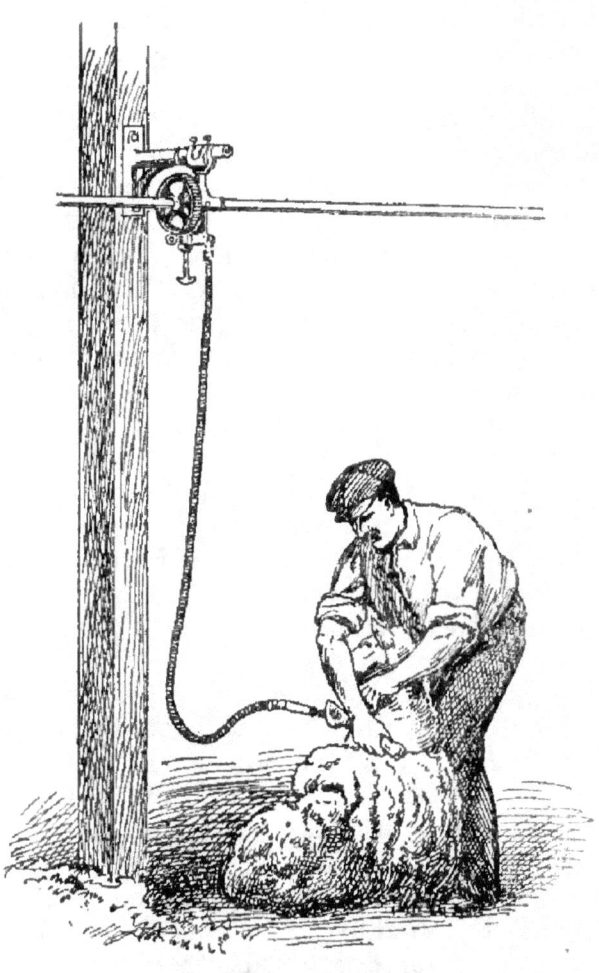

SHEARING BY MACHINERY.

The skin may be further tightened by stretching the shorn portion towards the shearer; but if the uncut wool is pushed back small pieces of skin are lifted up and are inevitably cut off by the shears. The shearer learns by practice how to make portions he is working on bulge out, using his own legs and body for the purpose. In holding the sheep he often has to hold a portion of it between his arms and body, leaving both hands free to manipulate the skin and the shears. The cutting should not be done with the point of the shears, but with the portion two inches or more below. If done with the point there is great liability to raze the skin, causing it to show a red mark, and rendering it very easy of attack by flies. The shears should always be kept flat on the part to be shorn, with the points slightly raised. The lower blade should be employed as a guide through the wool to open it up and to make a definite shearing line; for this reason it must be kept steady and the upper blade be brought down to it. This distinguishes shearing from clipping. Clipping off locks is of a similar nature to clipping a hedge, where both blades are moved. A clipping action is distinctly opposed to low cutting, as the shears very soon " ride out " of the wool. The point of the shears should be thrust into the wool, and then the hand be allowed to fall back slightly; this permits locks which have been over-pulled so as to lift small pieces of skin into a dangerous position to tighten on the body, so that, being then in a natural position, they are not cut. The operation is unconsciously performed after a little practice, and it is of the greatest importance where the wool is matted together and where the skin is thinnest.

The shearing should be done in such a manner that the lines are perfectly vertical, as then the neatest appearance is obtained; and it is highly essential that the wool be well cleared off the back along the spine. If the wool forms a ridge along the spine the sheep's back appears narrow. Everyone knows the advantage of a flat back as representing good condition. Very skilled shearers can make the left and right hand cuts meet exactly on the spine. The beginner should not aim at taking a wide cut, and the neatest work is done with a " blow " from half to three-quarters of an inch in width. Should the skin be cut, a small quantity of tar should be put on the wound to keep off flies, and to keep out injurious matter, for occasionally blood-poisoning is caused when the wounds are left unprotected.

Care should be taken to keep the fleece clean and unbroken. It should be carefully wound into a bundle. It should be laid out flat, the inside downwards, then the sides turned over so that it forms a narrow strip eighteen inches or so wide; beginning at the tail end it should be compactly rolled up, and by twisting a portion of the wool about the neck to form a bend, be tightly bound. Wool should be stored in wool sheets in a dry, sweet room, and it is essential that it be perfectly dry when stored. A damp floor or damp walls must be avoided. Sheep carrying very short broken wool may not be possible to shear with a whole fleece, in which case the fleece requires to be tied by string.

When shearing stock sheep less care is necessary than when fat sheep are to be sent to the market at once, when the greatest neatness is desirable. In such cases the shearers take as broad blows with the shears as the farmer will let them ; and the sheep present a very rough appearance at times. In taking a broad blow more wool is left on, and as in such work regard is not paid to the hang of the wool, it is not cut so short as it might be. In Scotland a considerable portion of the shearing is done while the men are on their knees, only the opening up about the neck being done while standing. This is suitable for small mountain sheep, but the strong, heavy sheep in England are more conveniently held in the manner already described. In Suffolk most of the sheep are shorn by hand, and are placed on a slightly raised platform, the legs being strapped ; this entails a cramping of the sheep's body without any great gain over the free method more commonly adopted with other big polled breeds. It is probably a continuation of the practice followed where the Norfolk and Suffolk sheep were horned ; and certainly there is advantage in having horned sheep strapped, as the horns present inconvenience to the free use of the arms and body in holding the sheep.

CHAPTER XIX.

DAILY MANAGEMENT OF LAMBS AND TEGS.

While with their Dams.—Having given an outline of the management of the lambs and tegs throughout the year, the details of the work among them from day to day may be dealt with.

As soon as the lambs get strong on their legs, the ewes go out to feed on grass or roots, taking their young with them. The lambs eat nothing at first, and should be provided with shelter in the form of hurdles, set up to break the wind. While very young, if the weather is wet, it is well to make a covering shelter against rain, similar to that in the lambing yard, as lambs, though little injured by dry cold, are easily chilled by wet. In the course of two or three weeks they will begin to nibble at green food, therefore they should be allowed to run forward on the freshest and tenderest keep ; at the same time they should be tempted to eat corn. Nothing is better than pea husk, with a small quantity of linseed cake broken finely ; the dust from the broken cake should be sifted out for this purpose. Very little of this should be placed in a small trough in the lamb pen, near to the lamb hurdle or gate provided for their exit. They should be taken back to the fold every night.

When they get stronger, the quantity of corn should be increased, a small quantity being given whenever they have cleared up the previous supply, but no stale corn should be left in the troughs. A small quantity of hay should also be allowed. It is more economical to feed hay from sheep racks than to let them eat it on the ground, when a portion is sure to be trodden in and wasted. By the time they are two months old, they require feeding systematically, as they have by this time acquired large appetites, and must be regarded as little sheep, still receiving a share of their support from their mothers, but becoming less dependent upon the latter. Half the corn at first given to the ewes may, by this time, be given to the lambs, and it may gradually be withdrawn until the ewes receive none. In seasons of short keep, it may be necessary to give both ewes and lambs corn to eke out the food If carefully fed, and no special disease breaks out, no serious trouble need be apprehended. The lambs require, however, to be kept free from lameness, therefore the feet should be kept sound and in good shape. Too strong caustic frequently contorts the feet, and renders the animal more likely to contract lameness at all times through its life. The paring of the feet should be restricted to keeping them in a natural shape, and as the hoof grows quickly, unless the ground is dry there is great likelihood of its not being worn down sufficiently fast to keep it in shape. If the lambs are caught once a week, and any abnormal growths are cut back, they will rarely suffer. However, in wet weather, especially if the lambs are kept on the dung in the lambing-pen, there is likelihood of the skin between the claws being abraded, and lameness, at first from soreness, and subsequently from foot-rot, will develop. I repeat, a mild caustic, such as the solution of sulphate of copper and vinegar, mentioned previously, should be used. If this is done at the first appearance of lameness, foot-rot will rarely establish itself.

On grass, lambs get little help in the shape of green food beyond what they get from the pasture, and if the pasture is good they have a good diet. It is, however, beneficial for them to have other food, and if kale sprouts or finely-sliced mangels are brought to them daily, they will profit thereby. Later, the autumn-sown catch crops are available, and a short time each day spent on these will be advantageous; corn, of course, being given as before mentioned. The early maturity of Hampshire lambs is in no small way induced by the frequent changes of food supplied to them during a day, and this is evidence that lambs require a mixture to thrive to their utmost. Hampshire lambs change their feeding ground as much as four or five times a day on those farms where ram breeding is a feature, and growth and condition are essential to the success of their sale. Folds are set on vetches, cabbages, in the water meadows, on rape and temporary pasture, in which the lambs get the first run over, spending a few hours in each daily. This system is the most thorough practised, and, of course, need not be followed in its entirety where the most rapid feeding is not required. Under ordinary circumstances, lambs at three months old do well on two changes a day, and they do not then require to go to the lambing-fold for shelter. If vetches, trifolium, or other catch crops are available, it is an advantage for them to spend a few hours there, whether they are on grass or on temporary pasture during the remainder of the day, though, of course, many do well folded entirely on good clover, to which a small quantity of other green food is taken. But it is undoubtedly an advantage to have some change in diet. As the ewes' milk falls short, it is important that the lambs be supplied with water. If this is kept constantly by them they never drink to excess, though when given occasionally there is danger of it. A block of rock-salt should be kept in the fold, so that they may lick it as Nature dictates.

It is always recognised that lambs, in fact, all sheep, do best where not in too large a number. This is particularly the case where lambs are receiving corn or special food, as the stronger shove aside the weaker. The weaker should, therefore, be kept by themselves where practicable.

In Summer.—At weaning, the lambs should be taken a long distance from their mothers, so that they will not hear the call of the latter. Their food should be particularly good on weaning, so that they may not miss the effect of their mothers' milk. Their folds should be changed daily, so that they get a fresh feed. The dipping should not be delayed, as the wool is long enough to carry sufficient of the solution to keep them free from ticks, lice, and scab through summer. As they are small and have little wool, the operation is quickly effected. An autumn dipping is also beneficial, and the lambs will be more comfortable and thrive better for it in the spring. Properly constructed dipping troughs are supplied by makers of the better kinds of dipping powder, and these are more convenient than a rough arrangement of tubs. It is always advisable to allow the sheep to drain on a piece of fallow ground, so that the drippings of the solution do not affect pasturage which will be eaten. The sheep may be fed through the hurdles meantime.

The dipping has a very marked effect in keeping off the fly which causes maggot on sheep ; but after a time it cannot be relied upon. Throughout summer one of the shepherd's first duties is to detect sheep which have been struck with the fly. The fly generally settles on moist parts about the tail, though the loin and shoulder are favourite positions. A keen shepherd will notice the discolouration of the wool very early, while a careless man will let the maggots commit serious pain and injury before he can see that the animal has been struck. Early signs are that the sheep is inclined to draw away from its companions, and that it switches its tail frequently. Later, it looks round at the place, and perhaps pulls the wool. Then the wool becomes loose and broken, and a streak of brown, foul moisture is seen. In large fields, with overgrown fences, sheep are occasionally lost in the surrounding ditches, through the carelessness of the shepherd, and are eaten alive. Mercurial ointments were commonly used a few years ago to get rid of the maggots, but they were dangerous as, if too freely used on broken skin, they were liable to cause poisoning. Simple remedies are found in milk and turpentine, and many other mixtures, but McDougall's Fly Oil is by far the best I have used, for it not only at once kills the maggots, but heals the wound. Before disturbing the colony of maggots, it is well to put a little of the mixture round it, to prevent their crawling outside the point to be dressed. If in a position where a blemish will not show, it is well to cut off a little wool above, then to pour in some of the mixture, and shake out the maggots. It is advisable to sprinkle a little flowers of sulphur on the wool about the place, to prevent any fresh attack. In wet, " muggy " weather, when the sheep-fly is especially active, it is a good plan to choose a time when the fleece is wet to pen the sheep closely and well powder them with sulphur by the aid of a powdering bellows.

If lambs scour they are liable to be struck about the tail ; to prevent attack, sulphur should be sprinkled on with a dredger. Trouble is saved if the wool about the tails of all the sheep is trimmed to prevent the accumulation of dirt down the thighs. Nothing is more unseemly at any period of the year than filthy locks from the tail to the hocks. An old pair of shears should always be at hand to keep these parts clean. The shepherd has a few special opportunities of discovering if any of the sheep under his care are out of health. These are when he first arrives in the morning, and when he gives them fresh food. His first duty is to look out to see if any of the sheep are at a distance from their companions, as this is a suspicious symptom. Such sheep should be closely noticed. Again, when fresh food is placed before them, whether in a new pen or in the trough, those which are failing linger behind It may be taken that whenever a sheep does not take to its food there is something seriously amiss with it, and that it will soon recover or soon die. Sheep medicines of a reliable nature were few until recent years, but great advance has been made since it has been recognised that farmers appreciated reliable medicines, and were prepared to pay for them. Modern veterinary surgeons make a strong feature of sheep diseases, and supply good remedies ; and the medicines supplied by several firms owning proprietary articles have been of great value to the flockmaster. There is no excuse for a shepherd who allows a sheep to die when he has once noticed any ailment. A sheep which dies a natural death is not worth a shilling more than its skin. A shepherd should always possess a good knife with a long blade, suitable for cutting a sheep's throat, bleeding a sheep, and for paring feet. It is important that troughs in which corn is given to sheep shall be moved daily, so that the sheep will not unduly manure one portion of a field to the detriment of others. This refers to pasture as well as arable land, and to all seasons. It is of highest importance in winter time when feeding off roots which will be followed by a barley crop. It is necessary that the whole crop of barley shall be uniform, and if the sheep are brought to one place with undue frequency this place will produce a rank growth, which, in all probability, will be stormbroken, or at any rate, produce corn which will cause the sample of grain to be uneven. The folds should, therefore, be set regularly, and similar quantities of food be fed in each.

The tegs which are shorn in summer time should be shorn in time to allow the fleece to grow again before winter, or they will not do well, as the wet will go through to the skin. Under natural conditions the wool rarely gets wet down to the skin, as it lies in a manner which conducts the rain outwards. If this wool is very short in winter the water runs straight through it, and the sheep, being chilled, cannot thrive. Ewe tegs which are not to be sold fat, but are to go into the flock, are not often shorn. Shepherding is easier in summer when tegs are shorn, as fly are easily detected, and clagging (trimming off foul and dirty locks) is not necessary. With few exceptions, shearing is a mistake where the young sheep are to be kept through winter, though those which will be killed before December often benefit by it. When once the natural thatch of wool is destroyed by shearing, water runs into the wool instead of off it, with the consequence that in rainy winters the skin, and therefore the body, is constantly wet, and the sheep exist in a state of chill.

In Autumn.—In autumn, when the sheep come on to the roots, the system of shepherding alters, as there is more close folding. During the day the sheep may have a run on the stubbles and young seeds, or grass, but at night they preferably go on to cabbages, rape, or white turnips. The folds should be made sufficiently large for one day's feeding, but a fair amount of room should be left for them to fall back upon, as in wet weather the pens get muddy, and on most land afford bad lair ; the food also gets muddy. It may be taken as sound practice in autumn and winter shepherding that a good large fall-back is necessary, and that in summer it is advisable. During the early part of autumn the turnip cutter is not required, as the food is soft ; but later on, when the swedes are used, it is undoubtedly an advantage to cut the roots. The extra expense of cutting involves the necessity of getting up the swedes and cleaning them ; however, the whole of the crop is eaten, and the sheep thrive far better. Against this the cutting is saved ; but a portion of the turnip has to be pecked up or it is lost. It at any rate gets dirty, and even ewes are damaged, as gnawing causes the teeth to get broken before they would otherwise be. A shepherd should be able to attend to 200 fattening sheep, close folded, all roots being sliced by him, if the turnips are got up into heaps made a chain square apart. There are thus ten heaps on an acre, each heap containing two tons. Allowing 1 cwt. per sheep per week, a heap would carry 280 sheep one day ; or 200 bigger sheep eating 22 lb. per day. The turnips are thrown together as pulled, except that if they are to be stored for some time before being used the tops are cut off. The shepherd should clean these, cut them, and feed 200 sheep per day. In addition he should move and reset the hurdles as wanted, fetch the cake, corn, chaff, and hay, keep the sheep's feet sound, remove boulders of dirt from their bodies, and keep them clean behind. It is also expected of him that he will dig out the patches of couch or twitch in front of the sheep. When no cutting is done he should be able to shepherd 400 sheep. Unless the shepherd has sufficient hurdles his work is very much cramped, as he may not be able to leave a proper fall-back, or he may not have sufficient to make a fresh pen before breaking up the old one.

One of the shepherd's chief troubles is to *keep the sheep within bounds.* On open plains and heath he relies very much on his dog during the day, bringing the sheep to the fold at night. On mountains, where the range is extensive, they roam very much at will, being overlooked in a general way, but shepherding as understood in enclosed districts is little followed. Almost all kinds of fencing are used to keep them in check, hurdles, either wattled or slatted, being most commonly used for close folding on roots, where by their close confinement they are most likely to exert themselves with the view of getting out. Wire-netting and string-netting, although used for close folding, are more suitable for dividing fields and other large areas; nevertheless, in districts where hurdles are difficult to obtain, they are of great service. Wattled hurdles are specially useful in exposed situations, as they break the force of the wind, and afford some shelter from the sun in hot weather. Sheep are liable to lie too much under them, and thus make an uneven manuring. Sheep are bad friends to growing hedges, as they eat the young shoots at the bottom, and thus weaken them in their most important part.

Hay is often fed to sheep through hurdles. This puts an extra strain on the hurdles, and is wasteful of hay. Hay-racks should be used. The simplest are those used in Wilts and Hants; they consist of a narrow longitudinal frame into which spars are fixed from side to side so as to give a trough-like appearance. These are filled with hay, and then turned over on the flat side, when all the hay is eaten without a portion being trampled into the soil. Heavy racks are inconvenient to move in the field, whilst these are easy to move and simple to mend. Wooden troughs are preferable to iron on account of the weight, whilst zinc and other metal troughs soon become battered in the rough usage they get. The simplest form is the ordinary pig trough or V-shape, though those made with a bottom board are most capacious, and are better suited for containing chaff and corn as well as roots. An advantage of metal troughs made with semicircular bottoms is that they are

L

very easily cleaned out. The troughs should be 7 ft. or 8 ft. in
length.

In Winter.—When the sheep are entirely fed on roots they
require sufficient troughs for all to feed at one time. First thing
in the morning the shepherd should clean out the troughs, and
give the sheep their chaff and corn. If the sheep have hay instead
of chaff it can be placed in the hay racks. If chaff is given they
are induced to eat more of it if they receive meal well mixed with
it, which is often desirable, particularly when first attempting
to make them eat a considerable quantity of dry food. The value
of getting sheep to eat dry fodder in early autumn is great, not
merely because it is useful at the time, but because, should snow
fall heavily, or through the muddy condition of the arable land,
they have to be removed, and put on to dry food, they will take

SHEEP RACK FOR HAY AND CORN.

freely to it ; whereas those which have not been accustomed to it,
will be slow to take it, and will sometimes almost starve before
doing so. For the same reason it is advantageous to break them
early to cake or corn. When giving the corn the shepherd should
make his dog keep back the sheep until all the corn is in the troughs,
so that all may receive an equal share. This is especially necessary
when sheep are being fed at high pressure ; for sometimes sheep
become cake greedy, and, being quick eaters, will eat far more
than their share, with the result that they overdo themselves,
and often die, whilst the slow feeders do not get their share. He
should carefully watch them to see that they come up promptly,
and that when they have got to the troughs they feed, otherwise he
may miss seeing an ailing sheep. As soon as the corn is cleared
up he should fill the troughs with cut roots until he notices they
begin to draw away to rest, and digest their meal. He may then

get his own breakfast. After this he may clean another supply, or set another pen, according to necessity. Before turning the sheep into a fresh pen he should go over it with a four-tine fork and dig out any small pieces of twitch which may be there.

If any sheep show signs of lameness they should be caught and dressed. Very often lameness is caused by dirt, or stalks of turnip leaf, which should be taken out, and a mild caustic wiped in between the claws. This will, in most instances, prevent foot-rot. In case of foot-rot the sheep should be isolated and kept from other sheep until all traces of disease are destroyed. Whenever a sheep is dressed for lameness, a temporary mark of ochre should be put on it so that it may be recognised easily. Before going to his dinner the shepherd should give the sheep another feed of roots. A good opportunity is generally afforded about dinner-time for the shepherd to put up the cake, hay and chaff that he will require for the following day. In the afternoon he can continue his work of cleaning a fresh supply of roots, moving hurdles, or other necessary work, and can give his sheep their second supply of corn about an hour before he will leave the pen at night. When the corn is cleared up he can commence to give the sheep their supper. He should continue this until all the sheep are satisfied, then he should fill up the troughs and leave them filled. The sheep will then rest contentedly through the night.

The last duty before going away is to carefully look round the sheep, and notice that there are none ailing. Then all the hurdles should be tested to see that they and the stakes are safely secured. After a drought, when it is difficult to drive the stakes far into the ground, the change to heavy rains occasions risks of the hurdles becoming insecure—in fact, they will sometimes fall almost of their own accord on account of the loose state of the soil. This occurs also at the break-up of a frost. Special care is, therefore, needed on these occasions. In winter, at the approach of frost, when severe weather may be expected during the night, the stakes should be loosened, or in all probability they will be frozen into the ground ; and if the frost continues the hurdles will be rendered useless until after the frost breaks up. In all long frosts large numbers of hurdles are frozen in this way, and the penning of the sheep becomes a matter of difficulty.

Fold-Setting.—The shepherd has less hurdle-setting to perform if he takes two pens instead of a single pen down the field at once, as there are then only three sides instead of four sides to set, the middle row forming an outside to both pens. On page 144 the usual plan of folding sheep is shown. The full lines show the outsides of the folds ; the dotted lines indicate where the dividing hurdles have been, but are removed to form a fall-back. The line A shows the row of hurdles which is left behind when only a single pen is worked through the field ; this line becomes available on taking the next pen down the field. When setting hurdles the stakes should be placed on the outer side, as greater resistance is obtained

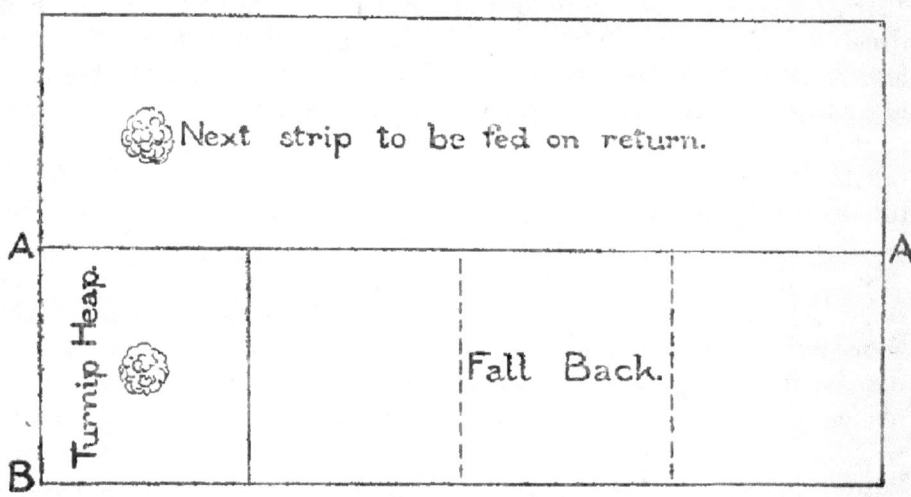

SHEEPFOLD, SHOWING FALL BACK.

so, and the shackles should be held tightly by the stake, as when loose the stake stands alone, but when tight it forms one of a continuous range, each one giving support to the next.

When using the ordinary slatted hurdle with a stout iron shackle to hold the fold stake, it is most convenient to drive the stakes into the ground by means of a heavy iron beetle, made cup-shape

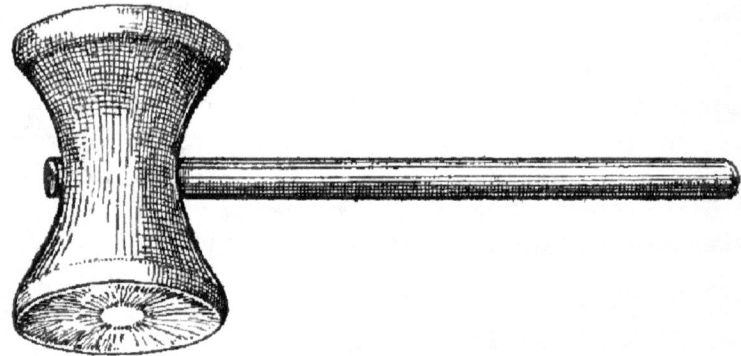

A CUPPED IRON STAKE BEETLE.

at each end so as to prevent the stake from splitting by constant hammering. However, in exceptionally dry or frosty weather a special stake, shod with iron, called a drift stake, must be driven into the ground by means of a beetle to make a hole in which to place the fold stake. The heads of the hurdles should be driven into the ground as additional support. Where wattled hurdles are used it is a common practice to use a withy (twisted hazel or willow), or a stout hempen shackle, which is slipped over the top of the stake and twisted round the end upright spars of both hurdles, and so made tight. In this case the hole is made by an iron crowbar or fold bar. Light "flitting" hurdles— those easy to flit or move easily—are best suited to work in a root crop for close folding, as by help of

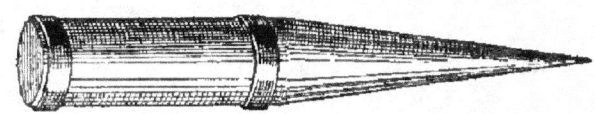

IRON-SHOD DRIFT STAKE.

the shackle and stake they are not merely coupled together better, but, by proper coupling, a considerable number of hurdles adjoining will be helped. These flitting hurdles depending mainly on the stake and shackles, therefore, have comparatively short feet to the heads.

In a few localities very heavy oak hurdles almost as strong as gates are used, the heads being made exceptionally long to admit of their being driven into the ground so as not to require the support of extra stakes, and an iron fold bar being used to let the heads into the ground. Although the hurdles are substantial

IRON FOLD BAR.

and last for a long time, they are not so profitable as the lighter forms, as they are expensive to move. There is no doubt that where used these hurdles are not moved as frequently as is desirable, and they are not met with in the strictly close folding districts, being more suited to use where little moving is required. Wire netting or hempen netting is extensively used for folding sheep, especially in districts where suitable wood for hurdles is not readily available. There are many points in favour of their use, but in the closest folding districts there is an opinion that wooden hurdles better sustain heavy pressure should the sheep, from any cause, such as fear, throw special strain on the fencing in the endeavour to stampede.

Inequalities in the ground in a sheepfold are dangerous, as sheep are liable to become "*cast.*" A sheep is cast when it gets on its back and cannot get on its feet again. Unless the wool is long and the sheep fat this rarely happens on level ground ; but sheep with long wool, particularly in warm weather in spring, when the ticks become irritable and the sheep roll to get relief, very frequently get cast, and, if not set on their feet, choke. When the wind blows into their mouths, they choke very quickly. On grass-land it is necessary to be among long-wool sheep constantly in the spring or heavy loss may be sustained. In-lamb ewes are liable to displace the fœtus and lamb prematurely if cast. The danger of casting is one reason why sheep lying in a big field should be put up by the sheep-dog when the shepherd goes to them, as it is possible for him in a hurried glance to overlook one, which would die if not turned on to its feet. The shepherd should see that the turnips are well cleaned before being placed in the turnip cutter for slicing. A piece of sickle or fagging hook makes the best cleaner. When a shepherd receives no help he must set the turnip cutter near to the heap of clean roots and fill the machine by means of a spike driven into a short stick, held in one hand, while he turns the machine with the other. It is quicker and easier than filling the hopper and then turning the machine. Kohl-rabi are very hard, and the root-stalk must be cut off; a skilful man cuts these off as he chops the root with the long-handled adze usually employed for getting up the crop, but a careless man leaves on a stump which is so tough as to make the slicing hard work.

So much of the profit of sheep-keeping is derived from the *manure* that the management of the sheep must always be carried on with a view to distributing the manure so as to do the greatest good and the least harm. When feeding on pasture the troughs should be moved daily and not be kept near the gate to save trouble to a lazy shepherd. The sheep follow the troughs, and the manure is thus more equally distributed. The same applies to all cases where sheep lie out in large fields. Special pains must be taken on arable land where it is intended to take a corn crop subsequently. The troughs require constant moving, but another matter requires attention. When turnips have been heaped and the rootlets and dirt require cutting off before they are put into the turnip cutter, a heap of rich mould accumulates, and this, if not spread carefully about the ground, will cause a gross growth on the spot, probably resulting in the crop falling before harvest and becoming stained. These small patches of fallen grain are very difficult to keep separate, yet in the case of barley they will cause such an uneven sample that the corn will not realise so much by several shillings per quarter throughout the field. Dirt and other accumulations in the troughs should also be spread about and not allowed to lie in a heap. For the same reason when sheep are placed in a small pen for handling they should not be allowed to remain too long, or the patch will become over-manured.

Unhealthy Foods.—A shepherd of experience generally knows to what he should attribute any ordinary ailment. The matter of paralysis through giving the sheep an over-supply of nitrogenous matter has been gone into in a previous chapter, and need not be enlarged upon here. If the *bowels* are too much *relaxed* it is either due to too much corn, too many roots in an unripe, rotten or other unhealthy condition, or to something injurious in the hay or chaff, though an internal chill may cause it. The farmer and shepherd should at once look into the matter, determine the cause, and rectify the diet. If the roots are rotten or unripe, while the other food is good, the cause may be expected there, and some portion should be withdrawn, and the deficiency made good by bulky dry food, such as hay or chaff. If the roots are good it generally points to something wrong with the corn or cake, which may have been heated at some time, or become mouldy. This is frequently the cause, and when it is, requires correction. Mouldy hay may cause it. We had a strange personal experience of poisoning on one occasion when the whole of the ewe flock was affected. The ewes were suckling lambs, which were fortunately old enough to eat food in addition. The sheep suddenly lost the use of their hindquarters, the milk dried up, and they scoured badly, the fleeces being covered with coppery scum and no little mucus. There appeared every likelihood of their dying. The shepherd, finding them in this state, gave them more chaff, which, fortunately, they were unable to eat. It appeared that the shepherd, when on the way to the hay stack, passed by where a stack of wheat was being threshed, and noticing what he thought was an agreeable smell from the chaff blown out of the machine, thought it would be an excellent food for the sheep. The sheep ate it freely, with the result mentioned. The sheep were given castor-oil, and then fed on rich linseed cake, split peas and sweet hay, and recovered, not one being lost ; about half the sheep could get up in the course of two days, and all of them within a week. The lambs were shut away from their mothers, and suffered very little, the drying up of the milk probably saving them. The milk gradually came back. The injurious matter in the chaff consisted of the seed-heads of stinking camomile (mayweed) and wild marigold.

Mayweed often causes *sore noses* to sheep feeding on stubbles. The stubble pricks the noses of the sheep, and poison gets into them, causing festering sores and (sometimes) gums. Such sheep are best isolated and fed on easily eaten foods. A contagious form of sore nose is much more to be dreaded, especially where sheep gnaw roots, as the contagion is caught as the sheep go from root to root. Sore noses and mouths prevent the animals from getting a proper amount of food, and they lose condition, the effect on a whole flock being a very serious loss. Sore noses, from whatever cause originating, should be regarded with suspicion, and sheep affected should be withdrawn from the flock and isolated at once. Bad cases should be bathed, and cooling ointment applied The farmer should handle a few of the sheep at least once a week to see whether they are in an improving or a retrogressive condition, in order that he may regulate the food accordingly. It is generally sufficient to run them into a small pen, and handle them on the loin and dock. A man accustomed to handling sheep soon finds out whether the sheep are doing well or the reverse, and young farmers should practise this frequently to acquire the knowledge. The condition of the sheep indicates whether it is advisable to hasten them on to the butcher. Local trade for a special class of meat often regulates the time at which sheep should be got out, though, of course, the general market has to be regarded also. The amount of keep the farmer has at command must also be considered. If there is a likelihood of its running short, the obvious course is to feed the sheep in such a way as will fatten them while it holds out.

Selling Out Tegs.—Fat tegs are most generally sold off in small quantities as they become fit for killing. Store tegs are oftener sold in large numbers. It is frequently found advisable to divide the tegs in two or more lots, so that they may be hastened on as appears desirable. This may cause a little extra trouble to the shepherd, but it is well warranted. When sending fat sheep to market it is usual to sell them in small groups. They undoubtedly sell better when each pen is made up level by selection. Those of the same size, colour, type of head and condition are best drawn together, as they look better and sell better so. Even when selling larger numbers they should be made to match. It is wise to draft out those which differ considerably in size, quality and type, and sell these separately. It may be taken as a safe rule that inferior sheep spoil the appearance of the better, and that the better do not improve the appearance of the poorer to a corresponding degree. If the sheep are sold in the wool, trimming off loose wool about the face, dirty locks of wool about the body, and a squaring of the tail always make them more " matchy " and more saleable ; a rough ungroomed horse never sells as well as one smartly turned out, and this holds equally good with sheep. Fat tegs which are *shorn* immediately before being sold should be done neatly and skilfully. Where the work is done carelessly, particularly if the back is not neatly finished, giving it a ridged instead of a flat appearance, the animal is placed in a prejudiced position, as it looks thinner and narrower than it really is. The legs and head also should be neatly trimmed. An extra shilling paid for shearing a score of sheep well is money judiciously spent, as each sheep may fetch from a shilling to two shillings more. If branded with a pitch mark before going to market, a small neat brand should be used. A large brand carelessly put on detracts from the appearance. If fat tegs are shorn in cold weather, they should be kept warm, otherwise the meat becomes chilled and will not set properly. A supply of thick jackets, made on the principle of a horse rug, should be kept for the purpose.

CHAPTER XX.

SHOW SHEEP.

Management of a Show Flock.—The general principles in the management of a show flock do not differ greatly from those adopted in the case of any other well-managed flock, but greater care is exercised in details. Show flocks are as a rule essentially breeding flocks, as the expenses connected with exhibiting cannot be met unless other portions of the flock are sold at an enhanced price on account of the reputation made by those which have incurred extra expense in their preparation for exhibition. Exhibiting is, in fact, the most approved method of advertising the merits of a flock. It is therefore largely in the matter of getting up for exhibition that the difference between an ordinary well-managed flock and a show flock exists. This is not entirely the case, however, as the most valuable portion of the flock for sale and exhibition consists of the rams.

It is useless, of course, to attempt breeding for exhibition unless good stock ewes and rams are kept. A change of blood has to be introduced from time to time to maintain vigour and improve the type; this is most economically done through the rams, as, though high-priced individually, they influence a large section of the flock. The occasional purchase of a few ewes from a specially good flock is desirable, as they are useful for mating with the rams of the home flock. The selection of the ewes is an important matter, as they are the permanent section of the flock, and are representative of its type and capabilities. The breeder's watch-word must be "Improve," and he must aim at a type, and breed and select to that type. The rams must be purchased with that view, and must be selected to supply deficiencies and alter the character of features in accordance with the standard determined upon. The soil has a great influence on the type, some kinds of land tending to the production of coarseness, others to too much fineness; and the variation is often observable on farms lying very near to one another. One farm is favourable to the development of specially good rams, while another produces ewes of great merit. Where the ewes run fine a stronger type of ram is generally necessary, and where coarse a finer-bred ram. Where ewes are finely bred, in-breeding is not so advisable as where there is a tendency to coarseness.   In-breeding to a slight extent is generally productive of smaller and more finely bred animals.   Carried beyond a certain point, the sheep become so fine as to indicate weakness.

The ewe sections of show flocks are usually rather better fed than those kept for ordinary purposes, as the aim is generally to produce sheep of good size, although not at the expense of quality. Smallness does not necessarily indicate fineness of quality, and sheep may be coarsely bred though small.   Small joints are cut from small sheep, but they may lack the quality of those from larger sheep.   The ewe lambs and theaves, until they come into the flock, are usually kept in a fairly fresh condition.   It is, however, the ram section that receives the most liberal treatment, and this is most marked in the case of the Hampshire sheep, of which a large proportion are used when lambs.   In such instances the ram lambs have the first picking of all the best food, and all other sections of the flock are of secondary consideration, as they are put with the ewes when they are only eight months old.   The ram lambs are fed as much corn as they will eat from the time when they first feed until they are sold.

In those flocks where the rams are not used until they are shearlings it is not usual to feed them at such high pressure.   A liberal allowance of corn is given while they are lambs, and this is continued to keep them in a thriving and growing condition until a few months before the sale, when they are fed on as heavy an allowance of corn as is consistent with safety, the object being to get them thoroughly fattened.   It is found necessary by breeders to fatten their sheep more than is consistent with the activity and vigour looked for in a sheep going to service, because buyers will not recognise the valuable points of the animals unless they are developed.   High feeding tends to the production of better wool, the appearance of finer quality of meat, a better outline, and more level handling, and shows definitely how much flesh the animal can carry.   If these points are not developed buyers will not give credit for them, and the fault of over-feeding is one which is rendered necessary by the purchaser, and should not be ascribed to the feeder who does it on compulsion.   Ewes too highly fed lose the power of breeding, consequently many breeders are very chary of showing ewes ; naturally, to be successful, the best ewes have to undergo the process of over-fattening ; and as they, or such a large percentage of them, are spoiled for breeding purposes, it is obvious that yearly drafts in this way greatly hinder the improvement of the flock.

House-Feeding.—Although at one time it was strongly urged that the most profitable management of sheep was associated with house-feeding, the practice has not extended very largely, nor is it likely to. The cost of bringing bulky food to the sheep precludes it, except in special circumstances, such as the preparation of sheep for exhibition, or the production of "house-lamb." House-feeding tends to the improvement of the quality of the wool, and the warmer atmosphere doubtless has an effect in making the animals fatten more quickly. The wool is also kept cleaner than when the sheep are fed in muddy folds; consequently a large proportion of the sheep exhibited are kept in the house. The sheephouse should be well ventilated, and the floor kept dry and well littered with dry straw, otherwise foot-rot will break out. House lamb is not so commonly produced as it was, as it is found the lambs thrive sufficiently well under ordinary circumstances. It has generally been restricted to those which are born in autumn, and are required to be killed at Christmas or soon after.

A TRIMMING RACK WITH ADJUSTABLE BARS.

Trimming for Shows.—The preparation for showing is an important matter. It is usual in the case of short-wool sheep to cut the wool on the back, so as to leave a square level surface, and also to cut in on other parts to give a good outline. Long-wool sheep are less clipped into shape, the object in their case being less to show the frame than the length and quality of the wool. In preparing the short-wool sheep it is necessary to begin to level the wool several weeks before exhibiting, and to go over the work again at intervals. Under the regulations of most Agricultural Societies sheep above one year old must have been shorn within some fixed period, generally April 1, previously to being exhibited, otherwise the wool would be left on from year to year, and would not be truly indicative of the produce of the sheep under the conditions in which they are generally kept; there would also be a want of uniformity in appearance which would be misleading to those not well acquainted with sheep. In the eye of a good judge old wool is distinctly detrimental to the appearance of the sheep, as it lacks the bright lustre and freshness of newly-grown wool, and a man of even little experience can at once detect whether the sheep was shorn bare at the assigned time.

The sheep to be trimmed is placed in a rack specially made to hold it by the neck (in the clutches shown on the rack in illustration), in a convenient position for the operator. It is first sprinkled with water, and the wool is cut off the back to give it a level appearance. The wool is scraped with a wool sorter's comb, in order to break the locks and bring up the longer hairs, so that they may be trimmed off. The brisket is trimmed to give a bold, deep front. From the brisket to the neck a full outline is aimed at, the endeavour being to make an even curve, showing an unbroken sweep from the jaws to the brisket, in the same way that from the top of the shoulders to the poll the upper curve is made to show a natural sweep with no inequalities. The hind quarters are cut square, and to effect this it is generally necessary to leave a greater quantity of wool about the thighs than about the hips. The tail must be cut square ; and though a full broad dock should be aimed at, it should not be allowed to hang so far back that it appears like an eave overhanging the rest of the hind quarters. The loose locks about the head and legs require clipping, and in most breeds the loose bunches are pulled out at the first trimming. Trimming in this manner is performed on three or four occasions, so that the wool forms a dense pile. An experienced judge can tell how much the sheep is trimmed into shape by the varying appearance of the wool, as the lower it is cut on any particular part, the finer and closer it appears. No colouring or other treatment will deceive a competent judge, as the lie of the wool indicates its length ; though as a "flat-catcher" to the inexperienced it may serve its purpose. To the experienced it reveals at once the slackness and want of correct outline in different parts of the sheep which a touch of the fingers will confirm. Good trimming, like good grooming, adds to the appearance of the animal, but its chief value rests in the favourable impression it makes at first sight, thereby attracting attention.

Colouring for Exhibition.—It is the custom among exhibitors of several breeds to colour artificially the outside of the wool. Red or yellow ochre, mixed in oil, is commonly used, according to the fancy of the owner. Other breeds are merely oiled. Colouring with ochre is not a practice of recent introduction, as it was commonly practised in Hampshire more than a century ago. Bachelor, in his "Cursory Notes on Husbandry," mentions it as a common and old practice there more than a century and a-quarter ago. It is, therefore, not the modern innovation it is often supposed to be. Its object is to give uniformity of appearance to a number of sheep together. The wool is so much influenced by the colour of the soil on which the sheep have been folded that when several lots from differently coloured soils are brought together, some of them will appear of a more favourable colour, and will consequently give an impression of greater value than others, although they are in reality no better. Colouring the wool is, therefore, a justifiable practice so long as it is not done to mislead. Where, however, colouring is applied to hide faults in breeding, it becomes reprehensible. For instance, dyeing the hair on the head, ears, or legs of sheep, which are not true in colour and marking, in order to make them appear so, cannot be too strongly censured, as also should be the eradication of rudimentary horns, because these practices are performed to hide blemishes in breeding, and to give the sheep credit for typical characteristics which they do not possess. The leading societies rightly make offences of this kind the ground for disqualification of exhibits.

Oiling or greasing the fleece is carried to an absurd point in some long-woolled breeds. It may give a suggestion of body to a fleece and help the lustre; but where all do it there does not seem to be much advantage gained; though again, in moderation, it may be useful to bring all to one colour.

The ochre is applied in the form of powder, or as a thin paste. When the colour is required to last more permanently, it is made into paste with oil, a small quantity being smeared on to the wool, and then rubbed on by hand. When the colour is not required to stand so long, a time when the wool is wet from light rain or heavy dew should be chosen; the powdered ochre should be sprinkled evenly over the sheep through a flour-dredger, and then be worked in by the hand.

Sheep travelling to a show should be fed to a great extent on dry food during the previous day to prevent scouring. It is advisable also to protect them by means of sheets, so that the wool does not become dirty or broken.

Success in feeding sheep for shows depends on suitable feeding (according to breed). There is not much difference however in the food given, as is shown by the foods administered at shows. In most breeds the sheep are fed on as big a variety as is available, and as the sheep are fed at very high pressure, this is very necessary. The great point is to keep the concentrated foods well balanced, and to supply only the best.

CHAPTER XXI.

MANAGEMENT OF A BREEDING FLOCK ON THE CHALKS.

A description of the management of a mixed flock, where the sheep bred on the farm are fattened almost exclusively in winter on roots, has been given. It is now proposed to describe the management of a flock mainly kept for breeding purposes, and where few but the ewes and theaves are retained on the farm in winter to be fattened. No breed lends itself better to fattening out on roots than the Hampshire, and none is treated more systematically or on more definite lines as a breeding or wet flock than this breed, where kept on its native Down land. It is well suited, therefore, to illustrate this type of sheep management. Moreover, it shows what an influence catch cropping may have on sheep farming. After the sectioned description of the management of the general flock, it is not necessary to enter quite so much into detail, or there would be unnecessarily frequent repetition. The treatment in this case will, therefore, be more general.

In dealing with the management of a specific breed, it has to be borne in mind that what are known as show flocks, that is, those which are kept especially to provide the highest bred sheep to be sold as pedigree stock, are subjected to much more generous treatment than are those flocks where the exhibition of sheep at shows is not intended. In breeds like the Hampshire, which has been so long and so consistently well managed, few really indifferent sheep of the breed are met with in the native district, or on land adjoining of practically similar nature. Some distinction between these classes of sheep will therefore have to be observed while discussing the management.

Autumn and Winter Food Supplies.—In the description of sheep management previously given, it was seen that the greatest call on the food supplies was through the autumn and into winter; but with a breeding flock in which fattening is little done in winter, and where the lambs are sold away in late summer and early autumn, the lightest call on the food supply is made when the fattening flock needs most. In fact, the Down farm is run at its lightest by sheep from September until lambing commences; from lambing until the drafts of lambs are sent away is the time when most food is wanted, and the quantity required naturally increases daily as the lambs grow and require more. July with the breeding flock is a very different experience to that with the flock a few weeks later. As most farms depend greatly on the arable land for sheep food, the system of farming is very much guided by it. As a matter of fact, on the really good sheep farms the sheep dominate everything; and not infrequently the shepherd dominates the master, for the shepherd says when a crop shall be sown, how big an area he needs, and elects when he will feed it. This would not be understood in most districts, but, on the whole, it seems to answer well on the Down farms, where a race of very responsible shepherds, with a long ancestry of shepherds, is found. Of course, on land more difficult to farm, it would not do for the shepherd to, as it were, interfere with the cultivation in this manner, because troubles in working the land, and getting the crops in properly, would probably occur. But the chalk downs are the most simple of all soils to work. The soil is light, it is generally thin, so that deep ploughing cannot be done, and it works at any time when not frostbound or when it is not actually raining. The heavy cultivations and tillages associated with stronger soils are not needed, fallowing operations to drag out couch can be done even in winter, and owing to the presence of much lime, turnip and rape crops can be grown in frequent succession. So also can "seeds" crops, vetches or tares. Sainfoin holds the land several years, unless it be too frequently grown.

Rapid Cropping.—The cheapness of tillage operations has another effect—the failure of a crop to stand, as, for instance, turnips or rape. It is not regarded in the serious light that it would be where land works less easily, and where the cost is great. Moreover, from the ease of putting in a substituting crop after the failure, the loss of time is not so seriously regarded. The fact is, there is little regard for rotation; a rapid system of cropping, as exigencies demand, dominates; and, on the whole, on these light soils, such methods are the best. Some sort of skeleton of an idea of a rotation is hazily sketched out, but in froward seasons even this has to be set aside. Of course, the heaviness of the stocking of the land by sheep greatly influences the opportunities for maintaining a rotation, but even when a comparatively light head is kept, the rotation generally solves itself into procuring a certain number of corn crops in a given number of years, the fodder and forage crops meanwhile being juggled to meet the flock requirements. The light soils run quickly to a short, knotty twitch or couch; but should the flock call for keep at some time ahead, thorough cleaning is not always waited for, just enough is got off to allow the crop to produce the needed food at the time it is required, and cleaning is deferred. As a matter of fact, a great many successful farms on these thin soils are by no means kept absolutely clean, but the rubbish is kept down to such a point that the crop's prosperity is not interfered with. The custom of leaving sainfoin down several years, and of keeping seeds down more than a year in many instances, tends to keep the land rubbishy, and as the roots go down but a little way, this method of only partly cleaning is not so unsound as those who farm clean on stronger soils might think at first sight. The method of farming is one which produces the most food, and the frequent breaking up of the land, with a little cleaning each time, prevents the smothering that appears to be probable. The cropping is mainly a catch cropping one, and it is only in this way that sheep farming can be done as intensively as it is. Moreover, the soils have little power of retaining goodness, consequently a quick succession of restorative and conserving crops is necessary—idle land has no place in such circumstances.

Ellman's System Adapted.—A general idea of the nature of the farming is necessary for the management of the flock to be understood. The system of cropping on the Wilts, Hants and neighbouring downs, is in reality an adaptation of the catch-crop system inaugurated by Ellman, of Glynde, in Sussex, when he undertook the improvement of the Southdown sheep some 150 years ago. It was useless to improve the sheep unless suitable food were supplied, and his genius saw that a catch-crop system, such as he evolved, was the only way to procure suitable food at all seasons, and allow the ewes to drop their lambs early in the year. Very little alteration has been made since his earlier days ; the introduction of trifolium and mangels as farm crops being perhaps the most important, and allowing the fullest scope.

Apart from the arable cropping the quantity of grazing available has considerable significance. As a rule, the Down farms are associated with little ordinary dry lowland pasturage, but those farms which have a frontage to those rivers where the water-meadow system has been developed, are preferentially placed. The extent of Down pasturage, of course, is important. On some farms there is comparatively little, whilst on others the range is extensive. As a rule, the farms with water-meadow frontages are associated with the more prominent flocks, as they provide suitable food for lambs in early spring with definite certainty ; and it is on these farms that the ewes can be brought to lamb very early— at the beginning of January. Farms which have belts of light to medium arable land lying between the water-meadows and the chalk arable are also favourably placed, as on these mangels can be grown, and mangels play a big part in the food supply from March to July.

A Wiltshire Rotation.—There is a rotation described as the Wiltshire, which is somewhat representative of the system of cropping on the Western Downs, and it would be difficult to formulate a more representative one, though I must confess that during the several years I lived on the Wilts and Hants border, I rarely saw it carried out with any strictness. Even were it set out to follow it, the difficulty of getting root crops to establish themselves in dry years, and the disturbance to the food supply immediately and for some time ahead, come as disturbing elements; and, more generally, the cropping is taken in accordance with the frequency with which the land can reasonably be asked to grow corn crops, and the requirements of the flock. Seeds, for instance, are often taken with greater frequency than are set out; moreover, when one year's young seeds fail, the old ones are frequently left to stiffen the food supply in the next year. The growing of roots in one year, followed by roots again, can only be accomplished on land with much lime in its composition. Indeed, it may be said that the land is subjected to a whirlwind of cropping, with never a lull in the storm. A very essential crop, sainfoin, is not mentioned, but this does not come into rotation, and may remain down three to seven years, according to the ability of the land to carry it.

The following is the Wiltshire rotation referred to :—First year, winter rye, trifolium, winter barley or winter oats, sown in autumn, as catch crops fed in spring, followed same year by roots; second year, barley; third year, clover; fourth year, wheat; fifth year, winter vetches, followed by late roots for spring feed; sixth year, forward or early turnips; seventh year, wheat; eighth year, barley.

In the section on The Green Food Supply a summary of the succulent foods, and the seasons when they are available, are given. It is possible to get these over a very large portion of the country, particularly on the lighter and easier-working soils; but whilst all these may be got on the Downs and accompanying land at the seasons given, it would be unusual to find all in other districts, though more catch cropping is possible in many districts. There are exceptional crops in districts where the climate is exceptional as, for instance, on the Suffolk coast the winters are so mild that August transplanted cabbages stand over winter, and are often relied upon to give lamb food in March. Hearting cabbages, however, can rarely be depended upon after Christmas, but thousand-head kale comes to its best in spring before the sprouts burst into flower. In February, in Down-land, white turnip-tops —both from those sown early enough to bulb, and also from those sown thickly in the autumn—are very popular with freshly-lambed ewes. In fact, very considerable areas of the latter are commonly provided. Rape is a very popular crop on the Downs, and is cheaply grown. Its seeding takes place at any time from March to September. The earlier sown becomes fit to feed in late summer; sometimes when fed very early it is fed off again in autumn, and then left to provide spring food. This is mentioned to show how the cropping is expressed on these lime-sweet, easily-worked soils. Although good swedes are grown over a big acreage, they do not ordinarily run to the heavy crops met with on stronger soils in moister climates, but they are relied upon as a good standby in late winter and spring, before serious attack is made on the mangels. To the breeder of early lambs the mangel is the sheet-anchor in later spring; the Down men recognise this, because, in cold, backward springs, they are a reliable food when the catch crops, upon which so much reliance is laid, fail to come forward. Moreover, in April, May and June, no matter how good other keep may be, mangels are always helpful.

Early Maturity.—With these remarks as to the sources of food and environment, one may more readily, and without occupying too much space, treat with the sheep. The improvement in sheep during the past century has done away with the necessity for allowing sheep to live three or four years before going to the butcher. In the grass-land districts where the root crop is not sufficient to fatten the sheep on roots during winter, the lambing season commences late, and the sheep, to a large extent, are not fattened until the following year, when rich grass comes again. With arable-land sheep the maxim is " a sheep a year," whether they are born early or late. With the Hampshire Down it is often a case of a sheep in three-quarters of a year—shorn in July, and every appearance of a wether in November—except the teeth. The high pressure of the feeding and growth is in relation to the intensiveness of the cropping. All the Down sheep are not fed at this high pressure, some receive comparatively little concentrated food; they may be bred in March instead of January, and be sold at the later fairs, to go eastward to the swede-growing farms of Hunts, Beds and adjoining counties, to come out as a second or third string of fat sheep off the roots, meeting the mutton market just before grass-fed mutton is available. These sheep, as lambs, are regarded more in the light of converters of the food of the farm, whether it be of the arable land or the Down pasturage, which they deal with in a very economical and profitable manner. The flocks of the Western Downs are generally on a large scale, as are the farms on which they are kept. Their success is largely due to the opportunities which large farms afford in comparison with those which small holdings provide. Only men with large capital, scope and skill could have modelled and developed such a splendid breed as the Hampshire Down. Flocks of a thousand ewes upwards are by no means uncommon.

Culling and Drafting.—In good flocks, especially in what are regarded as improving flocks, that is, where careful selection is made to bring in only such young ewes as will be graded the flock higher, there are always "coming on" young females to make good the yearly cullings and older sheep got rid of. A flock of ewes, therefore, consists of sheep of four ages : (1) the new draft of two-tooth ewes or theaves ; (2) the four-tooths ; (3) the six-tooths ; (4) full-mouthed and older. The new draft is reckoned to be slightly more than one-fourth, so as to meet the losses which, in the ordinary course, will be met as it passes through the stages until —in its turn—it is sold out ; and in like manner the proportion, as the ages increase, lessen. What proportion will go right through is problematic, as the "luck" of different years varies very considerably—it may be two per cent. or ten per cent. Owing to the dry lair, and the large portion of the time spent on arable land, the losses from diseases of an epidemic or epizootic nature are less on land where sheep are kept thickly than on grass land which has become "sheep sick." The ploughing of the land, and thereby the burying of germs or organisms, greatly sweetens it for sheep and keeps them healthier.

The flock is generally made up in July, where lambing is expected in early January, and all should be sound in every respect, doubtful or known wrong ones having been culled. Where lambing will not take place until much later, there is no urgency in doing this, though as draft ones should not be wastefully kept, culling may be done at weaning-time. They should be sound on their feet, and have been recently dipped. They should be in fair condition, and improving in it when they are needed to go to the ram ; the fecundity of any breed will be increased when part of their food is in the form of sweet (not sheep-stale) pasture. It may be on good grass, sweet Down, or, should the farm not provide this, sweet leys. The other food may be such as the farm ordinarily supplies, and a run on rape, or on stubbles where there is some fallen grain, especially barley, has a stimulating effect. The extremes of fatness and poorness should be avoided. Where Down runs are available, the sheep may run entirely on them after they are tupped, though it is generally advantageous to bring them daily to the arable land to scavenge, and eat such food as is available as well as to bring manure on to the land.

At this time there may be some lambs still awaiting sale, and these will have the best food growing, as well as such corn as is thought desirable. The ewe lambs drafted with the view to going into the flock next year, will also have choice food ; though usually they get less corn than the ram or sale lambs, because they will have plenty of time to grow out before they go to the ram. Although ram lambs are very generally used in these districts, it is exceptional for ewe lambs to be bred from, and experience shows that in so growing a breed it is not desirable to tax them with the carrying of a lamb and its after support ; the fall of lambs itself is not satisfactory, nor is the milk supply good. Where lambs are bred from, the breeding should be delayed as long as possible, but this again throws the next year's breeding late.

Ewes during Gestation.—The general management of the ewes during gestation has been treated with in the section " A Year with the Ewes," and need not be enlarged upon here ; but the mixed cropping of the Down farms supplies a greater variety than is commonly found on other farms. The ewes require exercise, and should not be too closely folded, and this the Down farms provide admirably ; moreover, the dry lair greatly favours them. It does not hurt them to come into smaller folds at night, and they suitably scavenge after the younger sheep. As a rule, the Hampshires come to the lambing pen in good working condition, but I do not think, from my experience, that they throw as many lambs as so strong a breed should ; or perhaps it might be better put that not so many are weaned. There seemed to be a little too much disregard for infant life, and too little acknowledgment of the early deaths ; but, in spite of that, the fall is not heavy as in some breeds, and the suggestion was that a little better doing before going to the ram might increase it ; for, after all, the " luck " in respect to doubles has practically nothing to do with treatment after gestation commences, though it has much to do with the strength of the ewe and lambs in the lambing yard. It is noticeable also with the Southdowns, which lead a much harder life than is commonly recognised. The fall of lambs from Southdowns brought on to good pasturage before tupping, and generally well treated, is decidedly heavy, but the hill yield at weaning time is not. Moreover, I have seen the same in respect to the Hampshire. The gospel of one good lamb being better than two moderate ones, is often much overstated, though in a show flock it may have rather more to commend it. A good ewe should nurture two lambs well ; but if she is allowed to get low in condition a little while before she lambs, she will be hard tried to get milk for them.

Prior to Lambing.—The later stages before lambing should see the ewe in a vigorous condition. " I never had my ewes too fat at lambing yet ; and though I have them good then, they are always poor enough within a month when they have a couple of lambs pulling at each, no matter how well they are done," was the statement of my late shepherd, Meadows, of Hollesley, who probably has the finest ewe and lamb record of any man in the country, as the records of the Suffolk Society for shepherds' premiums during the past twenty-five years show. In one year with me he had 587 lambs weaned from 302 ewes, with the loss of one ewe from being cast a fortnight after lambing, and one empty or guest ewe. But he looked to seal the contract of fecundity by having the ewes in good condition at lambing time. " I do not care what you give me when the ewes go to ram if you give me a good run on an unsheeped aftermath on good pasture ; I like a bit of rape for them, and a run on stubbles, during the day ; but it is good grass that makes the fall of lambs," was his view as to getting twins ; but wherever he could—which was not difficult with the Suffolks—he bred from twin ewes.

Shelter.—The Hampshire lambing pen has been described, and there is nothing better organised than these West Country temporary lying-in hospitals for ewes. They are necessary on the exposed hills, at the early date when lambing is done on them, and they are in every way efficient and economical. These and the field shelters of straw-stuffed hurdles to protect the lambs when the ewes take them out on to roots, are models in the way of housing. They are as hygienic as can be provided, and with the antiseptic aids which all careful masters and shepherds now provide, the mortality in the lambing pen, both of ewes and lambs, has been decreased in a most marked degree.

High Pressure Feeding.—The Hampshire is a big sheep, and can stand heavy doing. It speaks well for the sheep and management that, when the lambs are once on the way, they are fed at high pressure with exceedingly few losses. The variety of food provided doubtless helps this, which, in itself, is a point in management. With the better-done flocks, cake, or cake and corn, are liberally used, 1 lb., 1½ lb., or even 2 lb. of cake and corn is given to ewes where it is desired to feed the lambs at high pressure through their mother, and the ewes stand these great weights without hurt. Linseed cake is justly prized but the wise, and the only wise, precaution where feeding so heavily, is that of using a considerable mixture of other concentrated foods, thereby avoiding the risk that a too-heavily nitrogenous food would bring about. Of course, the large quantity of nitrogen used to make milk is, in itself, a safeguard, and prevents the dangerous accumulation of nitrogen in the blood, that is such a big risk with fattening sheep at high pressure ; but with it all, good management is necessary. The fact that the ewes are having swedes and turnips containing mainly starchy food, is a great point, for they help to balance the dietary. All ewes are not fed at this high pressure, but the good results obtained show that much can be done if care is exercised.

Value of Water-Meadows.—The ewes are kept out of the enclosed yards as much as weather will allow, and the exercise is beneficial to them. During this time the ewes are mainly fed on swedes, turnips which have run to top, rape, occasionally kale later on, and, where provided, a first-run over " seeds " or leys of Italian rye-grass or trefoil. Those who have the advantage of water-meadows are very preferentially placed, and certainly so in respect to very early breeding, because they can be relied upon to find plenty of suitable food for both ewes and lambs from April 1, and, at a hard pinch, after a severe winter, they may be called a little earlier to give the lambs soft, succulent food that the arable cropping has failed to supply. The Hampshire breeder, when he can get the sheep to the water-meadows, considers that his real difficulties in respect to lamb food supply are over, as he looks to the sureness of the ordinary catch crops to fill the gap before the leys are available. Water-meadows make the spring management of the lambs comparatively easy, because, owing to irrigating powers, there is no question as to their failing. In winter the meadows under water are not affected by cold, because running water must be above freezing-point. Meadows which, in the system of alternating the drownings or floodings, are uncovered in frost may get a check, but the effect of frost is soon removed as their flooding comes. Those without water-meadows, of course, have to exercise more skill to assure food at that critical period between the clearing up of the swedes and the coming in quantity of fresh young food ; Italian rye-grass, or that and trefoil, best suit as an alternative to water-meadows, but as there is not the same reliability in them, the general food supply has to be more definitely secured. Naturally, many play for safety by not allowing the ewes to lamb so early as those who have the advantage of water-meadows.

The need for a large supply of food for lambs in April, where they are born in January, is obvious, for gradually they have been getting the habit of nibbling turnip-tops while running forward through the lamb creeps into fresh folds where their cake troughs are placed. By the time they are from ten weeks to twelve weeks old they have acquired considerable aptitude to feed independently of their mothers. They are fed less through their mothers, and the cake is gradually withdrawn from the ewes and given direct to the lambs, but the ewes are liberally fed on roots and hay. Every week sees great increase in their consuming powers, and they soon acquire the capacity of little sheep. Failure of the catch crops is, therefore, a serious matter. It is a feature of the Hampshire management when the lambs begin to feed independently to give them folding on a run on more than one kind of food, whether or no it includes help from the water-meadows. Although rye-grass, winter oats, and winter barley are not highly nutritive, they supply tender and easily digested young growth, eminently suitable to lambs, and they possess the value of coming early, therefore considerable breadths of these must be provided. Trifolium comes in May, to be followed by vetches or tares, and all the time mangels are giving valuable help to both ewes and lambs. In June a piece of autumn transplanted cabbages of the early sugar-loaf type, may be available ; and after them, the hardier oxhearts come in. But vetches and tares are a great mainstay, and in no part of the country is so large an area sown. Ordinary leys and sainfoin, however, are the stock pieces of keep in summer-time ; and the arrangement of feeding these crops is to some extent controlled by the nature of the season, and therefore of the supply, which is managed accordingly. But at all times the lambs are provided with a fresh feed, that is, food that has not carried sheep previously that year. In all systems of sheep management, in any part, this fresh, as opposed to stale, keep is absolutely essential.

Weaning.—The lambs are weaned, roughly, at three months ; but there is no definite period, and as weaning is not done at a fixed age, but those early and late born are taken from their mothers on the same day, there is considerable variation. If lambs are doing well, and the milk supply is falling off, and a considerable number of the earlier have ceased to suckle, a generally average suitable time may be considered to have been reached. If, through scour or other throw back, the lambs are not thriving well, it may be advisable to give them the advantage of a rather longer suckle. In exhibition flocks the ram lambs are not castrated until weaning time, or even later ; because it is desirable to give them an opportunity to show their character and features ; those fit to keep on for service are allowed to, and those with deficiencies are " altered." Ordinarily, lambs are better tailed and castrated when about a fortnight old.

The lambs, when on their own, are kept liberally supplied with cake or corn ; and probably no breed of sheep sends out its lambs with so much " bloom " on them as the Hampshire. From July to October the lambs are sold off ; and it is a feature of the Western sheep that a very large proportion of the sheep are sold at fairs ; large numbers being pitched at old-established fairs, though to a considerably smaller degree than in former days. The sales of ram lambs from the better-known flocks often take place on the farms, though many are sold at local fairs and markets, and no inconsiderable number go to distant counties for sale.

Ewes after Weaning.—The ewes, after weaning, are run on land where the lambs have had first feed, or where down pasturage is available spend part of their time there. Shearing, culling, and other items in sheep management are performed much as with other breeds.

The ewe lambs which are selected to go into the flock in the following year are kept in a thriving condition, but are not pushed hard, except in exceptional cases. As lambs, they receive fresh keep ; but after a year, when they have built up their constitutions, the food is not necessarily so choicely selected.

The lines of selection, and the good treatment to which this breed has been long subjected, make it a fast feeder, whether as a lamb or up to maturity—and maturity in this breed comes very early. Old terms of nomenclature which were used in the order of lamb, teg and wether, the latter after being shown as a two-tooth sheep, are very frequently squeezed into as short a period as nine months—a veritable triumph in sheep breeding.

CHAPTER XXII.

MANAGEMENT OF A BREED LARGELY ON GRASS, AND WITH FEW CATCH CROPS ON THE ARABLE LAND.

Lincoln.—The Lincoln is typical of the heavier white-faced long-wools originating on grass, and though it is well suited to grazing, it has the faculty of doing well on arable land crops. It is, in fact, well adapted to mixed farming of its native county and those adjoining, though, except for the fen land, it has a tendency to go northward rather than south. It does well on ordinary upland grass or rich fens. Much that goes to the management of Lincoln sheep is covered in the general description of the management of a mixed flock, commencing with "The Sections of the Flock." But the Lincoln passes more of its time on grass. The lambing is arranged later to suit the appearance of grass; fewer catch crops are available in the spring; and with later lambing, and having in view the great value of the fleece, fewer are sold out early in winter, and more are fattened out on grass as shearlings, many being sold at the spring fairs to go on to rich fen pastures for the latter purpose. The value of the Lincoln to cross with the bigger Down breeds has been demonstrated very clearly in recent years, and has been largely made use of. These crosses can be got out earlier. The extent of the crossing is very noticeable to those who remember the rare exceptions of thirty years ago.

Flushing the Ewes.—The selling of fat hoggs begins about February, and continues through spring, whilst the more backward ones are run on to grass and are sold at big weights in summer. It is customary to put the ram to ewes in the first or second week in October, though in show flocks the ram is entered in early August, to bring lambs at the beginning of the year, the thick of the fall coming in February. The ewes are usually "flushed" by being put on mustard or rape, but they are not as a rule run expensively between weaning and this period, generally having secondary keep, stubble runnings, and the following up of the lambs, or second feedings. Through the pregnancy period they make use of the stubbles, leys and general run of food, until December, when they are put on turnips, again generally following the fattening sheep, where, with the necessary dry food, they remain until a few weeks before lambing, when they go wholly or partly on the pastures, receiving trough food to get them into strong condition for lambing. At the approach of lambing they are brought up to the lambing pen, and as they lamb they go on to grass, getting mangels, with such mixed corn, cake and chaff as the owner thinks well.

With the lambing in March, as the lambs get forward enough to feed, the ewes go on to seed leys or good pasture. On the more arable lands the clover leys are much utilised. As with other lowland breeds, the lambs are generally allowed to run forward through lamb hurdles or creeps, to partake of trough food. Weaning takes place towards the end of July, when the lambs get the sweet food from the clover aftermaths ; it is customary to give a small allowance of cake or lamb food, to keep the lambs going on when weaned. In August the cabbages are available, and carry them until the turnips are ready and wintering begins. Lincoln men, recognising the value of the wool, dip both lambs and ewes a little time before weaning, and it is customary to dip the ewes again in November. If sheep with these heavy fleeces are not freed from insects there is likelihood of considerable loss, as when the warm weather comes, and the insects get active, the sheep are apt to roll in their endeavours to dislodge them ; in doing this they are liable to get on their backs or be cast, when, owing to the length and weight of wool, they cannot get back. There is no more anxious time than this, and the shepherds, especially on rough-surfaced fen pastures, have to be round their sheep at almost all hours of night as well as of day.

Draft Ewes.—Draft ewes from the flock are fattened on mustard and rape, or grass. It is a striking scene to see a flock of sheep eat its way into a fold of the big summer rape, which grows so vigorously in the fens, that it far overtops the hurdles, making it no easy matter to see the sheep before they have stripped the leaves. Succulent as the leaves may be, the sheep thrive faster when they get down to the stems. In bad times with the sheep-fly maggot, the shepherd needs to be very keen to notice when the sheep is first showing signs of maggot whilst they are in this dense food. The system of management is one which is singularly independent of catch cropping. The late lambing does not call for it, and the general supply of keep meets the needs of the flock. Tares or vetches, rye and other crops, relied upon so much with breeds further south, are only comparatively rarely grown. As the wool is so long and plentiful, dagging or clotting to remove dirt and foulness has to be kept up as required, and, before shearing, this is generally systematically done. With the higher-bred flocks, where lambing is done earlier, with the view of getting the lambs forward for shows and the early sheep sales, kale, kohlrabi, and cabbages are more used, and more rye and tares are grown, and a far more liberal feeding with cake and corn is provided.

CHAPTER XXIII.

MANAGEMENT OF A MARSHLAND BREED.

Kent or Romney Marsh.—The Kent or Romney Marsh sheep had for its foundation a very big, coarse, long-woolled sheep, indigenous to Romney Marsh, which had much in common with the old Lincoln, and other breeds of the marshes and fens, as distinguished from the long-wools of higher lying lowland pastures. There is, however, much that is distinct in the breed, and its mutton is superior to that of any long-woolled breed. The Romney Marsh is generally an exceedingly rich alluvial tract of soil of geologically recent formation; much of it having been reclaimed from the sea, and the general appearance of the Marsh is very suggestive of the Fenlands of Lincoln and adjoining counties. The Marsh extends from Hythe to the River Rother, about fourteen miles, and in its broadest part, from Dungeness to Appledore, is about ten miles. This district is largely given up to the maintenance of sheep, but the breed affects a far larger area. Formerly, the Isle of Sheppey carried a somewhat similar, but inferior class, of sheep, and the Sheppey men of more than a century ago had to go to the Marsh for their sheep to improve and maintain their type and character. Those interested in this breed cannot do better than read the excellent description of " The Management of the Marsh and Kentish Sheep," written by Bachelor about 140 years ago, though published rather later in his " Synopsis of Agriculture." They would be struck by the great similarity in the treatment then and now, except in those cases where the most advanced flocks are pioneering in the breed's improvement, and are represented in competitions in the showyards. Even in them the ground work of management remains very similar. The breed has considerably improved since Bachelor's day, and the sheep kept on the Sheppey are in line with those on the Marsh.

Flukey Land.—A feature of the Kentish breed is the very large numbers held by individual flockmasters, for in few big breeds of British sheep are there such big flocks; these flockmasters number their flocks in thousands, and five-figure flocks are to be met with, though this is possible only through the custom of wintering most of the sheep away from the Marshes. Until comparatively recent years the Marshes were very wet and humid, so much so that few flockmasters resided on the Marsh, as ague was so prevalent. Better drainage and improved management, however, have considerably altered the climate, and ague is almost unknown. Moreover, the land is sounder for sheep. Much of the water is brackish, and as salt water is inimical to the snails which carry the liver fluke at one period of this pest's existence, the Marsh is singularly free from liver rot. The climate is not so consistently dry as apparently many have assumed, and the rainfall is far heavier than on many parts of the East Coast and the South-east Midlands. The Marsh, however, lies open and bleak, and at times is swept by very cold easterly and northerly winds. The strong winds had an effect on the degree of admixture of Leicester with the indigenous breed a century or so ago, for it was found that although the breed was greatly improved in many respects, it lost some of its hardihood, and losses were heavy from the habit of the crossed sheep seeking shelter on the ditch sides, whereby many were drowned through getting into the dykes; or, as one old writer puts it, were blown into the dykes.

Breeds Used.—For a short time Leicester rams were freely used, and then entirely withheld; moreover, since that time there has been no admixture of blood, all improvement going on within the breed—through selection. The Kentish sheep, however, had one most valuable outside help—through the Cheviot. At the time previously to the advent of railways, large numbers of Cheviot sheep were brought from Northumberland to the Thames by boat; and, in the ordinary course, they occasionally met bad markets in London; in such instances, to avoid certain market and shipping dues, the shippers unloaded on the Kentish Coast, selling the sheep locally. In this way there was admixture with the uncouth sheep of that time, described long ago as having coarse heads, thick necks, long, stout limbs, broad feet, narrow chests, flat sides, and great bellies, fattening slowly, wethers being rarely fit for use until they had completed their third year.

Influence of the Cheviot.—The influence of the Cheviot is very clearly shown in the outward form of the sheep; the distinctive ruff or frill of wool on the upper part of the neck being very marked. Doubtless, too, the very distinctive quality of the mutton making it superior to that of any long-woolled breed is largely due to this. Those who have been accustomed to the finer quality muttons find some of the long-woolled mutton insipid, and the fat too tallowy; but so tender and well flavoured is the Romney mutton, and free from the deficiencies of the coarser-fleshed long-wools, that it is popular in the markets, and realises a relatively high price. The Romney Marsh sheep, however, is not a product wholly of the Marshes, for although a greater portion is born on them, it is customary, and is a very old custom, to send the sheep on to the uplands of Sussex, Surrey, and more distant counties, to winter on pastures of an entirely different character. Many leave the Marshes on September 1 and return on April 1. In fact, on those dates the roads leaving Marshland and passing through the Weald are alive with flocks in migration. Only a very light stocking is left on the Marshes in winter. The Wealden soils are very variable, but the winter grazings are by no means confined to the Weald. Consequently, the influences of soil and climate, very different to those experienced on the Marsh, are brought to bear on the constitution of the animal ; so that the Romney, instead of being a narrowly-bred breed, having the indigenous character of one breed only, and being accustomed to only one narrow set of conditions, inherits primarily the Marsh features, with some assistance from the Leicester and Cheviot. It has been, through a very long period, influenced by wintering on the uplands. The conditions under which it lives are by no means easy at all times ; for though the marsh grazings are excellent, the winterings are generally done on hard fare, such hard fare as few big sheep would go through successfully. The features and constitution inherited through breeding and mode of living, befit the sheep to exist successfully under hard and very varied conditions, and since breeders have seriously and so successfully improved the type of their sheep, an extraordinary demand has sprung up for this great meat and mutton sheep. In the wet climate of New Zealand, in the harsh winters of Southern Patagonia, which is almost Canadian in its severity, and where the land is much covered with snow, and where strong winds blow almost unceasingly, in the wet humidity of the Falkland Islands, as well as in the more typically sheep-raising countries, this composite breed is able to put up a constitution which enables it to thrive as well as on its native ground. It is significant that more narrowly bred longwools, which were taken in hand earlier, and were brought to a high state of excellence much sooner, and greatly influenced the sheep industry of some of the great sheep-raising countries, are finding a very strong competitor in this comparatively new breed in exportation. From narrower breeding they do not so readily meet requirements where meat has an increasingly valuable ratio to wool, whereas, in these thinly-populated countries, before cold storage was available, the sheep had little value beyond the fleece. But as their wool-bearing properties were undoubtedly great, and they carried with them good constitutions, they exercised prepotency over other breeds. They were well suited to dry countries, and it must not be forgotten that they have profoundly influenced the sheep of the great wool-growing countries of the world. They will still be required to help in the work.

No Best Breed.—There is no best breed to suit all districts in this country; much more is there no breed which is best in all foreign countries, but the Romney is undoubtedly attracting great attention because of its combined wool and mutton properties.

The Romney is a perfect grazer; no matter what the size of the flock, or the dimensions of the pasturage, they intuitively spread themselves out with extraordinary evenness, and this is as noticeable when they are grazing rich grounds a dozen to the acre, or in winter, when there may be only one to the acre. No other breed ranges so perfectly, and this doubtless has an effect in keeping the pastures sweet, for the droppings are never massed together. When disturbed, however, they rush together very quickly, for, big as they are, they are quick travellers. On the road, they travel strongly and well.

Lambing in the Open.—Although some of the most improved flocks breed early, April is the chief lambing month. The ewes which have been agisting at a distance are brought back in time for this. The treatment at lambing is very simple. Severe weather is rarely experienced on the Marsh at this time of the year, and the lambing pen is practically ignored, in fact, often ridiculed as being unnecessary; though in one lambing season within the last four years, the loss of lambs was very heavy owing to a moderately severe fall of snow; which suggests that if the lambing pen is not elaborately set up, some shelter should be available in emergency. The ewes lamb in the open, which is sanitarily sound. Some breeders provide an occasional straw-stuffed hurdle, but the majority do little in this way. The lambs generally get well up on their legs in a short time. The ewes, however, are peculiar as lowland sheep, in that they do not trouble to have the lambs with them at all times. They graze away from the lambs very much, occasionally returning to them. They are, however, good mothers, supply milk plentifully, and are not really negligent of the lambs' welfare, though those accustomed to other breeds might be anxious at the apparent carelessness. Only in very exceptional circumstances is extra food given, but it is by no means difficult to get the ewes to take swedes or mangels.

Ewes Find for Themselves.—To a great extent the ewes may be said to find for themselves, and there is no more simple form of shepherding. This characteristic aids the breed when it is exported to the great ranching countries. Castration and tailing are done within a month of birth. Shearing is done in May and early June —the mechanical shearer being largely employed now. Owing to the heavy fleece, and in some instances somewhat heavy bellies, the Romney sheep require to be carefully watched, or losses through getting on their backs are frequent. Shearing time, therefore, brings a welcome relief to the shepherd. It is customary with many flockmasters to shear the lambs. This is a practice which is centuries old, and obtained when the wool was the asset to the sheep keeper. Little profit is made from the lamb's wool now. In fact, it is doubtful if the loss on the shearling fleece meets it, but shearing makes it more easy to keep the " fly " or sheep maggot in subjection, and those who are in the habit of agisting their sheep in autumn find that the losses from fly are greatly curtailed, especially as the sheep are often sent out to graze with those who are not necessarily skilled in shepherding ; and when the wool is long, they are liable to overlook attacks in the early stages.

Shearing Lambs.—But the shearing of lambs is not altogether beneficial. My experience certainly does not impress me favourably with it, even with this long-woolled breed. The shorter fleece matters little in cold, dry weather, but in prolonged wet winters, the sheep suffer badly from being soaked to the skin. Wool is a natural thatch when it has not been cut, but shearing very largely destroys the thatch principle ; for instead of the wet being led away from the skin, the rain goes straight on to it, of course, very severely chilling the animal. As it is wet rather than dry, cold winters that are trying to sheep, anything that tends to destroy the natural protection from wet, must further handicap the young sheep. Older sheep have gained a stronger constitution, and, therefore, shearing is not so prejudicial to them. The lambs remain a long while with the ewes, generally until the August sales, by which time they have attained considerable size, and have very big appetites.

Marshes too Wet for Lambs in Winter.—The Marshes are too wet for lambs in winter, whatever else may be left there. Most of the lambs go out to winter on the pastures ; but in superior flocks many lambs are wintered in the fold, commencing in the autumn on rape and mustard, following with turnips, and getting swedes after Christmas. The general custom, however, is to run the lambs in store condition, and they are generally sold as hoggets at two years old, though with considerable latitude on both sides of that time. The wintering of the lambs, ewes, and hoggets on grass is effected at little cost beyond the value of the wool ; this enables flockmasters to keep the sheep, as is often the case, over two winters, before sending them to the butcher—the wool being an important asset. The Wealden pastures are, therefore, practically a necessary adjunct to the Marshes. However, there is an increasing tendency to obtain swedes for sheep in wintering, now that the sheep are attaining more rapid maturing powers.

Changing a Breed's Character.—There is much modification of character going on in the Romney at the present time ; the more advanced can be brought to the butcher much earlier, where suitable food is available. How far it is advisable to alter the type is a point of some perplexity to the Marsh men. Is it advisable to alter the sheep much if the nature of the food available is not altered ? is a problem which has had to be answered in respect to many breeds. The solution has generally come by altering the food ; that is, by changing the systems of farming to accomplish it. The Kentish men are somewhat awkwardly placed, as, owing to the large proportion of heavy land in the area lying back from the Marshes, there is not much arable land suitable for root growing and winter folding interspersed among the pasture land ; and if the sheep are not wintered farther afield than has been customary, where they can get on to light land and more roots, much care will have to be exercised in alteration in the type of sheep. Considerable alteration, giving earlier maturity, has been effected on the advanced flocks, and the influence of rams from these is felt on the general flocks, with appreciable benefit in appearance and quality. At the same time, by no means all of the big flock masters are wholly in favour in the change of type brought about, as some of them do not think that they winter better on the food available. The tendency, however, is to bring modern modifications to bear ; and probably the situation will be met by giving the sheep extra food beyond what they find on the winter pasturing. This will mean additional cost ; but, on the whole, the breed can afford it ; for a very general price for wintering is only 5s. per score per week. It is the low cost of wintering that is very attractive to many ; but it is difficult to see how change can be avoided.

A Grazier's Breed.—Many graziers prefer that the sheep do not have roots, as they find that the sheep thrive better on their return to the Marshes in April when they have had nothing but grass. Moreover, many like the sheep to go back to them in a very lean condition. But on many points there is by no means unanimous agreement. Some like them lean and some otherwise ; and, on the whole, there is probably as much difference of opinion on points of management in this breed as in any British sheep. But, admittedly, the race is singular in many respects ; further, it is rather in transition, and these naturally make for variable opinions.

In handling the breed a little way from the Marsh, I have been surprised at its adaptability. That it will do well on even only moderate pasture is certain ; if it can get enough grass to eat it will thrive. Its natural food is grass, and it seems to want nothing more. Yet I find it a good sheep on roots alone, or preferably on roots, with a run at grass. I have been seriously informed by considerable flockmasters that ewes and lambs will not eat tares or vetches, and that if they ate them they would suffer ; my experience has been much to the contrary. There is no doubt it is essentially a pasturage sheep, and is most profitable where grass is the main food ; but it is quite certain that the sheep is susceptible to very great modification in its treatment should it be found necessary.

Crosses with the Romney.—The Romney crosses well with Down breeds, and the favourite local cross is with the Southdown. The cross, of course, brings about earlier maturity, and without injuring its grazing capacity, though it adds to its folding value. The mutton from the Down cross is very popular with butchers and consumers. Surprise is often expressed that such big mutton should sell well in the Southdown's country. The chief reason for this is that the Romney and Romney cross mutton is very tender, and can be eaten directly after slaughter, whereas the Southdown needs to be hung some time, because it is relatively hard. In hot weather, meat is naturally liable to go bad if long hung, and the Southdown goes relatively out of favour in the hot season ; though judges of mutton would naturally prefer well-hung Southdown. Much of the Romney mutton is consumed in the popular seaside resorts of Kent and Sussex, and as the visitors are not all of a class that is expert in the quality of meat, it secures a good market ; for its tenderness, flavour, and juiciness combine to make it acceptable to those who, in ordinary circumstances, eat little but cold-storage meat. It is quite certain that the mutton of the Romney sheep, pure or crossed, gains greatly in public favour.

Fat Lamb Crosses.—It is commonly accepted that the bigger Downs, such as the Hampshire and Suffolk, are most useful for producing fat-lamb. If the big Down cross lambs are not sold fat, but run on as stores to be sold in autumn, they do not attract the grazier, and sell relatively badly ; whilst the Southdown cross sells very well. In fact, the bigger crosses, when fattened as tegs or wethers, do not sell really well in comparison with the merit of the meat, when compared with the Southdown, the Romney, or the Southdown and Romney cross ; how far this is controlled by the butcher or the public is not quite clear. The chief selling periods of stores are from weaning time until October, and again in March, April, and May. Markets and fairs are very frequent during these periods.

CHAPTER XXIV.

MANAGEMENT OF A HILL FLOCK OF BLACK-FACED SCOTCH SHEEP.

I am indebted to Mr. Anderson, late of Longformacus, for the following description of management, he having kindly supplied me with the results of his experience when farming on the Scotch Hills, an experience which has not fallen to my lot.

We will take for example a typical flock of Black-faces on the Southern Highlands of Scotland, at an elevation of 500 to 1,000 ft. above sea-level. The flock would number 400 ewes, all acclimatised, being bred on the ground and never removed until sold off as draft ewes at five years old. They would be divided in ages thus (that is, if you enter the farm on November 1, after five-year old ewes are sold off): eighty, four years old, eighty, three years old, eighty, two years old, eighty, one year old, and eighty ewe lambs or hoggs, at six months old. You would require at this time ten rams to turn out on November 15, with a few in reserve to turn out later on if any one failed, and in case the hoggs, or any of them should take the ram. This is prevented by "breeking" them. They are taken to a convenient pen, a square of old sacking about 9 in. square is sewn over their tails on three sides, leaving the bottom open; this keeps off the ram. After the rams have been withdrawn from the hill the hoggs are again taken to pen and the cloth removed. The reason of this procedure is to retain the hoggs on their native grazings, so as not to undo their acclimatisation; if once taken into a field of better pasture you spoil their hardihood. About this season, the shepherd on his daily rounds gathers the ewes into lots as he may find them, taking care to have at least one ram with each lot of ewes. Some shepherds feed the rams with a few oats carried round with them; this extra feeding keeps the rams active. A careful shepherd teaches the rams to come up to feed before turning them out, but usually the rams are left with the ewes to help themselves and rough it until the service is over.

Ewes in Winter.—Much depends on the way the rams are reared —whether house-wintered or out-wintered. The ewes, after the rams are withdrawn, want little attention during the winter, and should the winter be an open one, they will require nothing but the produce of their grazings. The shepherd, meanwhile, carefully watches the signs of the weather, and is always ready to take his flock into a shelter when a storm comes on, the shelter being usually a conveniently-placed plantation, or a "stell" built of dry stones in circular form, with a port or two for the sheep to enter. These, if well placed, may save many sheep in a sudden storm. A stack of bog or meadow hay is usually placed near in the autumn, so that the shepherd can help himself, and as he sees need. No sheep racks are provided, the usual plan being to lay the hay in small handfuls upon the snow, taking fresh ground every day until the storm is over. About April 15 the lambs may be expected to appear, and, in ordinary weather, the ewes are not removed or disturbed at all.

Lambing.—The shepherd will now have his hands full, and generally a lambing assistant is provided for him, say for six weeks, an experienced man at 30s. a week and board. They get out at daybreak, and walk the whole hill, which may be from 800 to 1,000 acres, as the usual practice is to allow two acres to one ewe kept. This is about as much as an ordinary grazing of bent grass and heather will keep all the year round. The shepherd on his rounds often meets with many difficulties during lambing—sometimes a ewe down, exhausted with heavy labour, or a lamb coming wrong ; sometimes a lamb found alone, and the mother to find—after finding, her legs must be tethered to prevent her straying again. Sometimes a dead lamb is found, and the ewe keeping guard. After examining her for milk, she is driven to a pen purposely prepared on the hill ; a foster lamb is found, often a twin, the skin of the dead lamb being drawn over the living, in which way the ewe is deceived. Some ewes take lambs without this trouble, but much depends on their milk and temper. All this is repeated often during the six weeks of lambing, exhausting the shepherd and his dogs. The dogs often fail before the man, and become quite done up before the end. After finishing lambing the castration of the male lambs takes place. The whole flock is gathered into a suitable suite of pens somewhere near the hill. The usual practice is for an experienced shepherd to draw the testicles with his teeth, chewing tobacco all the time as a disinfectant. Others use a pair of pincers to draw out the testicles, dipping them in carbolic as they go on. Another way, and a neater method, is to open the purse and press out the stones, clamp them, and burn them off with a hot iron, using a salve, and leaving the purse open. This entails no loss of blood. It is the general practice in Wales ; and the first-named in Scotland.

Washing and Shearing.—The shepherd will now have a quiet time until about shearing time, which usually takes place in July. A considerable amount of care is necessary just previous to shearing, as ewes get on their backs and die, or sometimes get their hind foot fast through their horn by scratching, and are found dead in this position. Washing preceeds shearing about eight or ten days. The usual practice is to take the sheep to a pool in a running stream and jump them in, repeating the process three times, not handling any of them. The water must be clean. Shearing is generally done on a stool of wood or of turf, the shearer sitting. The sheep is laid on its back until the belly is shorn, then the legs are tied with a cord or thong, and the sheep laid on its side and shorn from head to tail, or *vice versa*, finishing along the back-bone. The fleece is rolled up and tied with its neck or tail wool, and the sheep is pitch-marked before leaving the stool. After shearing, there is often much trouble to the shepherd in " mothering the lambs," and it often happens that some lambs never find their mothers and are then weaned. The weight of wool runs from 4 lb. to 5 lb. per fleece, lightest from the older ewes, and heaviest from the yearling ewes. Lambs are not shorn in a Black-faced flock. After shearing, little trouble is given the shepherd until weaning time, but in the meantime he is often busy cutting and drying " peats " for fuel, and making hay for a coming winter.

Weaning.—About September the lambs are weaned and sold direct off their hill The top wether lambs are usually sold to a butcher or wholesale dealer, killed and sent to our city markets. The seconds are usually sent to an auction sale, where they are bought by lowland farmers for feeding, coming to the butcher later on. The very top of the ewe lambs are kept for stock and run on with their dams, never being removed from the hill. The second ewe lambs are also sold to lowland farmers, and kept for stock purposes, often mated with Wensleydale and Border-Leicester rams, producing cross lambs of quick-fattening qualities and well-known feeders.

Drafting Flock.—About October the five-year-old ewes are drafted out. All with mouths and udders correct are sold in one lot, others with a tooth out or any other fault are sold as such, but a Black-faced ewe carries her teeth until she is seven or eight years old. The writer has often taken three crops of lambs from draft ewes,. afterwards selling them to the butcher, one very strong ewe nursing three pairs before being sold. The whole flock is usually dipped before the draft ewes are culled. Then again in about three weeks; but in all cases dipping should be done before the rams are let out, as much handling after is not good. An average crop of lambs would be 75 per cent., or 300 lambs. In a very favourable year, with a few twins, one may have about a lamb for each ewe wintered. From aged Black-faced ewes the writer has had as many as 50 per cent. twins, but these have to be better kept both before and after lambing.

Prices in 1913 for Black-faced sheep were : Draft ewes up to 32s., usually 22s. Top wether lambs, 18s. to 20s. Seconds, 10s. to 13s. Top ewe lambs acclimatised to the hill would be 30s. to 32s. Seconds sold off 12s. to 14s. Wool 6½d. to 8d. per lb.

CHAPTER XXV.

General Features in Sheep Management.

The diseases of sheep are treated subsequently by a veterinary expert whose qualifications and experience warrant the greatest confidence ; and with a preciseness and conciseness which, as a layman, I could not profess ; consequently, such remarks as may be made on the ailments here are merely such as arise in the daily course of management, and which assail the shepherd or flockmaster in the ordinary routine of farming ; and which influence the management of sheep in the ordinary course of their work. Many points have necessarily been dealt with in the foregoing pages, but there are some of these which can be more conveniently dealt with at greater length than is advisable in sections where a consecutive narrative of certain aspects in sheep farming is being set out.

The Sheep Farmer.—A good sheep farmer is generally a good all-round farmer, for no class of farm stock requires closer attention or more skilful management than sheep. Moreover, he must be a skilled tiller of the land, as the ability to secure suitable crops at all seasons proves he has his arable land well under control. When these are ensured, sheep farming is generally one of the most profitable sections of the farm. The sheep farmer must always look ahead well to provide crops for all divisions of his flock ; and in unfavourable seasons must be resourceful, so that if some portions of the food he might reasonably rely on prove to be failures, he may devise means which will enable him to carry his flock through a critical period. Such occasions occur frequently, and if the farmer has made no special provision, the sheep have to be sold in a bad market, and subsequently he has to buy in again on unfavourable terms. As a rule, the sheep management through-out a district varies very little in broad principles, although there is often considerable variation in the details. What appear to be minor details frequently make all the difference between profitable and unprofitable sheep-keeping. The customs of districts are generally sound, as they are the outcome of a long experience, which has shown what is best under the particular conditions of soil and climate ; it is, therefore, unwise to make profound changes unless all features of the changes are well considered previously, and are proved, at any rate on a small scale, to be more advantageous than those practices they are to usurp.

Knowledge of Markets.—Without change, however, improvements cannot be made, but alterations are not likely to prove valuable unless they are made by those who possess exceptional skill and experience. Those with great knowledge of the markets, and who are able to judge from the amount of food there is likely to be throughout the country at a given time, are able to foretell whether prices are likely to be high or low at a particular date in the future. These are justified in consuming their fodder at the customary period, or may reserve it so as to have it available at a time when sheep keep is likely to be scarce. For instance, in dry summers it is often difficult to get a good plant of turnips ; and even transplanted crops, such as kale and cabbages, which, as a rule, may be successfully planted at a much later date than is possible for swedes and mangolds to be sown to produce a remunerative crop, do not thrive. When this happens, the spring sheep feed throughout the country is necessarily small in quantity. Yet the same number of sheep have to be supported as in a more plentiful year. To keep the sheep healthy and thriving, the supply of spring food has to be drawn on, leaving little for that period known as " between roots and grass," which is usually from March to May, and depends very much in duration on the earliness of spring growth.

In seasons of great shortness it is often particularly remunerative to save the food until spring, and then to buy in sheep, which, as others cannot afford to keep them, are necessarily sold at reduced prices ; but when the fresh supply of summer keep becomes available they rise in value again, thus giving to persons in a position to purchase them in times of scarcity of food the advantage of the rise in price, in addition to the increased value due to their greater size and improved condition. When food is thus saved until spring, a much larger number of sheep can obviously be kept during the short period mentioned than if they had been kept throughout the winter, yet each individual sheep will probably pay more for the short time than those which were bought in the autumn. It is to an extent a speculation : so, however, is the purchasing of sheep in the autumn, when the risk of loss of life and disease has to be run for a long period. Many thousands of sheep are bought in in the autumn in such years, and the buyers know there is practically no chance of profit ; yet, because it is the custom of the district to buy at that season, they do it year after year. In no other business would such a course be followed.

Time to Buy.—The time to buy is when there is a fair chance of profit, and not when there is little likelihood of gain. The skill of the farmer comes in in buying at the right time ; he must watch the markets closely, and buy in accordance with the trade and the prospect of an early return of the time when keep will again be plentiful. If he misses his chance by being over-greedy, he may miss much of the profit he otherwise had it in his power to obtain. Late spring feeding of roots and other fodder crops tests the skill and resourcefulness of the farmer in the management of his land and cropping, as he throws an increased amount of work on his horses late in the season, and it will probably necessitate an extensive alteration in the crops he must sow. This is easier now than it was a few years ago, as few restrictions are made as to the order of cropping ; whereas, when narrow leases restricted farmers to a strict rotation, it was almost impracticable. It will therefore be seen that only those possessing special skill are justified in going outside the ordinary practices of the district.

Other illustrations might be given as to the stocking of the land in accordance with the prospect of ample or deficient keep at a particular future date, and of the consequent variation in prices, which allow the shrewd farmer an opportunity of obtaining an exceptional profit. They are by no means rare ; but the one given should be sufficient to illustrate the manner in which a shrewd man may obtain remuneration for the exercise of his greater skill ; and also to indicate to those less experienced that the subject of sheep management can only prove entirely remunerative when everything bearing upon it is thoroughly mastered.

The management of the food when grown so as to make the best of it, and so that the land may not be injured whilst it is being consumed, and the management of the sheep themselves, are dependent on the skill of the farmer, backed up by the skill of the shepherd.

The Shepherd.—The shepherd must possess an instinctive love and attachment to animals, and must be prepared to sacrifice his personal comfort and ease for the welfare of his flock, because in the experience of every shepherd there come times when extraordinary efforts have to be made, entailing strain on his endurance and a demand on his resources which fall to the lot of men in few positions. His is a position of great responsibility, and neglect on certain occasions would entail great losses to his employer. He must be quick to detect the slightest change in his sheep ; whether they are improving or going back, or whether there is any change in their demeanour, because, as sheep do not show their suffering until the cause has assumed a dangerous or exceedingly uncomfortable aspect, successful treatment depends much on its early recognition. There is nothing so important in the whole of shepherding as the possession of this faculty. That a shepherd knows his own sheep is to be expected of him, and to a commonly observant man is not extraordinary, having regard to the fact that he is daily among them and constantly handling them, although they may number some hundreds. He must know more. He must be so conversant with them that the slightest change in the demeanour of any one will be noticed at a glance. A really skilled man with sheep will know at sight if a strange sheep is out of sorts; though a sheep may have some misleading peculiarity that, until he has observed it for some time, he may not be sure whether it is really ailing or distressed, or merely possesses an eccentricity. Men not accustomed to sheep may walk through them time after time and not recognise ailments until they are very far advanced. A shepherd must be observant, or he is of little value except as a mechanical worker among sheep.

The Sheep Dog.—Moreover, he needs a good sheep dog, well trained, or his own labours will be greatly increased, and his time is well spent in training a capable dog. Sheep dogs can be trained to perform duties which seem to place that animal's intelligence above that of all other animals; and a shepherd without a good dog may be well described as only half a shepherd. The mere fact that there is a dog about sheep keeps them under control, but a good dog also directs and protects them. When lying wide he collects them; and, if properly trained, will go round hedgerows, where a sheep may be hung up in briars, or into gullies and other shelter, where those which are ailing are likely to creep. The ninety-and-nine in a flock can generally be found, but the hundredth should always be the shepherd's care; as whenever a sheep has anything the matter with it, its first instinct is to get away from its fellows. This is the case, whether it be from ordinary internal troubles, maggot caused by the fly, fly-troubles on the poll, bad cases of foot-rot; in fact, anything that is unusual. The shepherd should, therefore, take a tally of his sheep, and make himself sure that their number is correct every time he visits them. It is the first duty of an overlooker to be able to count accurately sheep lying out in a pasture, when at the troughs, when passing through gates, or from fold to fold, or in whatever situation he may find them.

Lameness.—Lameness should be noticed at once, because it is sure to need attention, whether it be from foot-rot, sore feet, dirt between the claws, foot-and-mouth disease, happily rare (though in the past two years I have three times been in a restricted area), strain, or a broken limb—the latter are important, for the inexperienced often handle a sheep in this condition with unintentional cruelty, the frequency of foot-lameness and the rarity of other sources of lameness, disarming the suggestion that the trouble arises elsewhere ; a simple fracture may become a compound one through wrong handling.

Catching and Turning a Sheep.—In catching a sheep in the open the crook is very convenient, and the hold should always be got at the hock, as, if taken lower, and the sheep swerves, the stifle may be put out ; this applies also to hand catching ; and if a tight grip is taken over the hamstring, the sheep soon succumbs. When turning a sheep on to its rump, the left hand should be placed well under the throat, the right knee low under the left flank, and the right arm well over and under the off flank, then with a quick heave on to the knee, the sheep will be jerked by the leverage, and easily set over. Very heavy sheep, such as big, fat rams, can be turned easily by gripping the under jaw by placing the right thumb in the mouth on the off side, then giving the head a sharp twist, and running the sheep back quickly whilst the neck is locked ; in the endeavour to avoid a crick, the sheep will sink on to its rump, and before it can right itself, a quick movement will enable the shepherd to secure it. This is not often practised, but at times it is very convenient, especially when dressing sheep for foot-rot, or to prevent it. Sheep are liable to get on their backs—to become cast—from which it is difficult for them to get back on to their feet, in which case death from starvation would be certain—but they quickly suffocate. The surface of land should, therefore, not be rough, as the risk then is greater ; deep furrows should be avoided, and ant-heaps and other lumps should be levelled. In view of the prevalence of foot-rot the shepherd should never neglect the feet.

Foot Troubles.—Ill-shaped feet conduce to foot-rot, because, sooner or later, some part of the foot will provide an easy inlet to the foot-rot germs. Although paring will not altogether prevent foot-rot, it minimises the risk very greatly. The best practice to learn how to dress a diseased foot is to work frequently on sound ones, and thus get an intimate idea of the form of the foot. With this knowledge a dresser can cut boldly on the worst misshapen foot without fear of cutting the tender portions ; whereas a novice niggles and frets away little piece after little piece, and probably ends up by cutting too deeply. When trimming a foot, it should be held up sole uppermost, to give the shape, and each claw should be trimmed to similar shape. A rather strong artery runs nearly to the end of the digit or toe—about half an inch back as a rule—and if this is cut there is profuse bleeding. To stop this bleeding the thumb should be placed firmly between the digits where they join ; by doing this, pressure is applied to the artery, and bleeding ceases. If it does not cease in a short time, a piece of wood, about half an inch thick, should be placed over the artery and be held down tightly by some binding. It may be necessary to leave it there a day or two in bad cases—to make a thorough heal. There is no advantage in making a foot bleed, or of applying a caustic so strong that the sheep is unduly pained. Yet one often hears shepherds boasting that they had done so. This shows a misconception of the nature of the disease. What is needed is to trace out every particle of disease by paring the hoof until at all points there is a clean junction of sound hoof where it joins the sensitive part of the foot ; the application of any antiseptic such as blue vitriol or carbolic-acid will then leave a clean and sound foot. Sometimes, however, the whole hoof may come off. Very bad feet should be put into foot-rot boots, or be wrapped round with sacking, so that dirt cannot enter the wounds for some days whilst new hoof is forming. When a shepherd finds sheep are falling with foot-rot, he should at once run them through a foot-rot bath containing a solution of sulphate of copper, give them a daily walk on hard roads to harden the feet, and in bad weather make them stand on lime for a time under cover. Of all things, he should avoid violent caustics such as butyr of antimony, except where there is proud flesh, as a distorted, hard hoof forms, which is very difficult to deal with should foot-rot break out again from a fresh attack, as it is quite possible for it to do.

On Flooding Land.—Sheep should not be allowed to graze on flooding land between June and the first sharp frost in autumn, for fear of their getting liver fluke, and at all times rock-salt should be kept before sheep. Husk, generally regarded as a serious ailment in calves and sheep, has ceased to be regarded so by me since I have adopted the practice of submitting them to chlorine fumes derived from bleaching powder on which a little hydrochloric acid is poured. If the sheep are placed in an air-tight building for a short time, and made to inhale the fumes, the husk worms are easily dislodged ; but the inexperienced must be careful not to let the sheep remain in more than a very short time, or they will suffocate. With this experience I do not worry about husky land. The circumstances are, however, different in the case of the husk which comes from lung worms, for they, like the liver fluke, when once they are established in the lungs or liver respectively cannot be destroyed without killing the animal.

Stale or sour green food has been dealt with already, and the losses from this are often very great. Land carrying cropping of this kind must not be confounded with land that is " sheep sick " or tired of sheep, as it is sometimes described. In the former case it is a question of the unhealthiness of the food which has grown up since a former crop was fed off by sheep quite recently.

Sheep-sick Land.—Sheep-sick land is that which has been overstocked with sheep, generally for a long period ; though on wet, heavy pasture land it may be for a shorter time, provided the sheep have been numerous, and a considerable number have been affected by ailments caused by organisms which give rise to some of the common, but very deadly, diseases to which sheep are heir. These, generally, are in the form of worms, or in a stage which, after entering the sheep, will develop into them ; though there are others such as are conveyed by other means, as through ticks and keds, which enter as specific diseases through inoculation. These and their treatment are dealt with in detail in another chapter ; but the importance of over-stocking cannot be too strongly impressed. Flock owners have, through long ages, recognised that if they sheeped their land too heavily they would be subject to great losses. They knew their land would become sheep sick, but they were not familiar with the organisms causing them, and which the microscope and closer observation have revealed. It is unfortunate that at present remedies have not been found for some of the most deadly diseases, and it is to be hoped that the extensive research work being carried out will shortly supply serviceable remedies and means of prevention. Good progress has been made, and if it rested only where the recognition of the diseases and their causes reaches, much good would have been done ; because already much that was obscure in sheep management has been made clear, and practices which were harmful have been stopped. Ordinarily, it may be said that sheep on pasture suffer less from stale keep than do those on artificially raised crops or arable land ; but that there is greater liability to sheep sickness on grass land because many of the diseases on arable land are greatly kept in check by the burial of organisms to a depth at which they cease to live, and because through repeated working of the land in fine, dry weather, they are denied the moisture which is essential to their existence. The thorough draining of arable land has made it healthier ; but grass land has been far less subjected to draining. The healthiness of the sheep on the light, chalk soils is well known, and no land carries such a heavy stock ; heavy, wet pasture land, if stocked nearly as heavily, would soon become sheep sick.

Driving Sheep Slowly.—Sheep should always be driven slowly, though light, mountain sheep seem to defy this axiom. The heavier breeds, however, are usually fed at higher pressure, and suffer when unduly rushed. It is particularly necessary not to wash or dip sheep directly they have had hard exercise and are heated ; they should be well rested and be allowed to cool down ; moreover, they should not be shorn when the bellies are full—they should have a partial fast previously.

Ear-Marking.—It is desirable to mark the sheep in a permanent manner, and wool marking by means of a pitch brand does not do this, although it is very useful as a temporary marking ; nor is pitch favoured by the wool buyer. The ear is the best place, and it lends itself to the purpose in several ways, such as tattooing, slitting, and punching. Tattoo punches prick through the skin, and leave an impression similar to the pattern of the punch, so that when coloured matter is rubbed in it becomes permanently fixed. Numbers, or any special design, can thus be placed on the sheep. Slits or nicks on the edge of the ear can be made to indicate anything, according to the code the farmer may adopt. A nick on a certain side of a particular ear may be made to indicate the ram by which the sheep was got, and the year it was got. As there are two ears with two edges, and as three or four nicks may be made on either side, there is sufficient scope to meet the necessities of a large flock. One or more holes may be punched in the ears to identify sheep. In pedigree flocks buttons with studs to keep them in place are used : these are placed in holes previously punched through the ear, and as any number or design can be stamped on them when being made, an easy method of keeping a record is obtained. The number should be recorded in the flock-book. In the case of pedigree flocks, the buttons should be made to close in such a way that they cannot be removed or used on a second occasion. A strip of soft metal is often used, the ends of which are inserted through two slits in the ear, and then turned so as to prevent slipping out. Numbers can be stamped on these to identify the sheep.

Sale and Purchase.—In previous chapters remarks have been made on points to be looked for in handling a sheep, and on the necessity of making even drafts when offering them for sale. There remain, however, a few points which may well be dealt with. So many animals now pass through the auctioneers' hands that there is little doubt farmers have lost some of the skill they possessed in estimating the weight of their sheep when they more frequently sold them to the butcher. Butchers realise that they can buy cheaper from the auctioneer than from shrewd owners, and naturally purchase from the former. On the other hand, the farmer is sure of prompt payment from the auctioneer, whereas the butcher was not always ready to pay on the deal. He, however, pays the auctioneer pretty heavily for the accommodation. It is a recognised axiom that animals always show themselves better on the farm than when tired and jaded in the market.

Selling is effected in several ways, by auction, upon legs, by live weight, or by dead weight. When sold *by auction* the farmer pays a percentage on the sale. " *Upon legs* " is a term used to denote they are bought as they stand, as when sold from the fold ; a lump sum being given for the sheep as they are, with no deductions or additions. When sold by *live weight* they are disposed of at so much per stone, gross weight. When sold by *dead weight* they are sold per stone dead-weight. It is usual to sell them *sinking the offal ;* this implies that the wool, skin or pelt, head, and the whole of the internal organs, with the exception of the kidneys and the kidney fat or suet, and the legs below the knees and hocks, are not weighed in, but a sum equivalent to their value is mentally calculated, and added to the price per stone of the whole carcass.

A considerable difference is, of course, made when sold carrying a heavy fleece or freshly shorn. A 12 lb. fleece, at 10d. per lb., naturally adds ten shillings to the value of a sheep, and if the sheep weighs 10 st., it makes a difference of a shilling per stone. The dead weight of an animal is taken when the body has cooled and dried. As the sheep are killed from home, it is necessary for the seller or his trusty agent to see them weighed. A simple brand should be put on the heads, and the head should not be severed from the carcass until the time of weighing, as there is no other way of identifying the body, and another may be substituted to the disadvantage of the seller. When buying by live weight, it is usual to calculate on the fasted live weight, as it is very different when calculated on an animal with its paunch full. A rough estimate is found on the basis of allowing 8 lb. of carcass and 6 lb. of offal to each stone of 14 lb. It is usual, therefore, to speak of a *stone live weight* as being 14 lb., and a *stone of mutton* as 8 lb., the two being described as *long* or *live weight,* and *short* or *dead-weight stones.* This is by no means applicable in all cases, as the percentage varies greatly between a store and an ordinary fat sheep, and, of course, far more in the case of a sheep fit for exhibition purposes. Condition and breed are matters to be considered.

Live Weight in Pounds.				Percentage of Mutton.		
		In Wool.			Newly Shorn.	
280 to 300	.. ..	61 to 72	.. ..		74 to 75	
260 „ 280	.. ..	69 „ 70	.. ..		73 „ 74	
240 „ 260	.. ..	67 „ 68	.. ..		71 „ 73	
220 „ 240	.. ..	65 „ 66	.. ..		69 „ 70	
200 „ 220	.. ..	63 „ 64	.. ..		67 „ 68	
180 „ 200	.. ..	61 „ 62	.. ..		65 „ 66	
160 „ 180	.. ..	59 „ 60	.. ..		64 „ 65	
140 „ 160	.. ..	58 „ 59	.. ..		63 „ 64	
120 „ 140	.. ..	56 „ 57	.. ..		62 „ 63	
100 „ 120	.. ..	55 „ 56	.. ..		60 „ 61	
80 „ 100	.. ..	53 „ 54	.. ..		58 „ 59	
60 „ 80	.. ..	50 „ 52	.. ..		56 „ 57	

Skins of sheep, whether with wool on or not, should be sold when fresh. It is a great mistake to leave them to spoil, as is so commonly done.

With something like 30,000,000 sheep in this country, and with no small portion of the country quite dependent upon sheep for its profitable utilisation, their successful management is an important matter. The value of a good shepherd has been mentioned, and the necessity for maintaining a supply of these skilled men is quite obvious. A shepherd is not like a mechanic, who can be trained under cover and made into a handy man in the shops. He must have a natural and instinctive love of animals; and experience has shown that an early training and association with sheep is almost essential, for few become really skilled in the art who have not taken up the work in their youth. It is unfortunate, but no less the fact, that the modern form of education does not encourage or befit boys for work on the farm; and that by keeping boys playing about on village greens instead of letting them get on to the farms among animals and the other matters of interest on the farm, at an age when they are receptive, is the great cause of migration from country districts. Country boys, with days broken by school hours, learn nothing of the pleasures and interests of country life; and by staying at school until they are fourteen they get a distaste to country affairs. The half-time system, with evening classes later on, turned out the best and most skilled workmen on the land. If this cannot be provided for all, there ought to be special provision for those who desire to take up shepherding; they are going to take up far more skilled and far more useful work than an ordinary clerk, and they need to start their technical education in it when the power of observation is keen on it, and before they are dulled to the appreciation of the attractions that country life affords. Those who enter their life's work young get an interest in life and in useful occupation, but the glamour of life in town appeals to those who have learned no real interest in country life—and they leave it at the first opportunity. It is to the nation's interest to make provision to maintain a race of shepherds, and it is necessary for educationists to recognise that work among animals is in itself a great education, and to do so without delay.

CHAPTER XXVI.

BRITISH WOOLS.

The tendency to produce crossbred wools in our Australian Colonies, to which special attention was drawn in the columns of *The Bradford Observer* some ten or fifteen years ago, is a tendency not confined to the Colonies—it must always have been a potent influence at home as well as abroad, and has undoubtedly affected the types of English wools produced in a very marked degree. The change in Australia from the small-bodied, fine-woolled Merinos to the larger-bodied, comparatively coarse-woolled crossbreds, was incident upon the development of the frozen mutton trade ; at home any such change is principally dependent upon wool values. When Blackface wool was down at 3d. to 4d., and other wools were depreciated in like proportion, the farmer's mind was a blank so far as wool was concerned. Sheep were bred for mutton only, and the wool was not regarded even as of secondary consideration. The wool factor among British sheep farmers would be of even less importance than it actually is, were it not for the importance which purchasers of sheep for export place upon stud sheep being well woolled. Purchasers of Romney Marsh stud sheep, for example, look for a sheep heavily fringed and with much leg-wool, whereas the Bradford wool merchant would prefer a fleece unfringed and clean cut, as, for example, in the case of the Border-Leicester. The American purchaser of stud sheep is obviously going to ignore no single factor, and he consequently takes carefully into consideration the wool factor, and gets every possible advantage. With wool at present-day prices, and in view of the possibility of a total reversing of marketing conditions incident upon under-production rather than over-production, the British farmer will do well to take into account wool values.

Two Types of British Wools.—The many and varied classes of British wools of to-day may most conveniently be studied under the two headings—Long-wools and Short- or Down-wools. The typical example of the Long-wool class is the wool upon the Lincoln or Leicester sheep, while the typical example of the Short- or Down-wool class is the wool upon the Southdown sheep. The differences between these two wools are very remarkable. The typical long-wool, as its name implies, is long—say 10 in. to 16 in. in staple—lustrous and wavy. The short-wool is comparatively short—say 4 in. to 6 in. in staple—comparatively non-lustrous, and "frizzy." As the uses to which these two typical wools may be put are very varied, and further, as the manipulative processes for the two are very different, it will obviously be advantageous to deal fully with each, and make these two typical styles a basis of comparison for the many British wools of more varied characteristics.

Before doing this, however, it is at least interesting to note that the usual explanation for the existence of these two types of sheep and wool in our island is that the rougher mountain breed—the progenitor of the Long-wool class—probably travelled across Europe with the wandering tribes striking north-westwards, and came into the British Isles from the north. On the other hand, the Merino sheep is supposed to have been developed upon the shores of the Mediterranean, to have reached its fullest development in Spain, and from thence to have been brought to the South of England and crossed with the native breeds, thus producing what are known as the Down or Short-wools. Although there seems to be a prejudice against the Merino cross in this country, it is more than probable that these are the facts of the case, for not only are there very suggestive records of such crossing, but in the experiments in sheep-breeding recently carried out on the Cambridge University Farm by Professor T. B. Wood, in which a Shropshire was mated with a Merino, there was marked segregation at the first cross, and this could only be due to either the Shropshire or Merino being impure, and as the Merino rams were specially selected, it seems more probable that the Shropshire has been at least originally impure, although, through the care of its breeders, it has now attained to a stability which practically gives it the right to a class name.

It is further interesting here to note that there still appears to be no evidence which suggests the impossibility of maintaining the Merino wool characteristics in this country, although the converse is frequently asserted. Curiously enough, the suggestion that the long-wool breeds came to this country from the North is indirectly given colour to by the researches of Mr. H. J. Elwes, F.R.S., who, in his many experiments conducted with a view of ascertaining the original progenitor of the sheep in these islands, has gone chiefly to the North for his primitive breeds, including therein the Shetland, Manx, Soay, Hebridean, the Black-faced, and the Orkney breeds of sheep. It is true that the wools of all these are not always characteristically long and lustrous, but probably few of these breeds are absolutely pure ; there are traditions, for example, respecting the Spanish sheep introduced from the wrecking of vessels of the Great Armada round the Scottish coast.

Typical Long-wools and Long-wool Breeds of Sheep.—Typical long wool, as required by the woolcomber, spinner, and manufacturer, should be long, lustrous, and wavy, of a uniform staple, and sound from root to tip. Fleeces should be free from black hairs, kemps, and bright hairs, and should be washed on the sheep's back, and marketed in this state, as the quality and lustre of long-wools may be much better judged in the washed state. The sheep-breeder wishing to make the most of his wools should carefully consider the above noted characteristics, for a very small divergence from type results in wool much less valuable. For example, if Lincoln sheep are not carefully bred, the wool has a tendency to deteriorate into straight, stringy, hairy wool, technically termed "britch." Thus, the Lincoln breeder endeavours to produce wools of a lustrous, wavy, fine, and uniform staple. But not only must he bear in mind the typical wool required, but he must also endeavour to produce a uniform fleece. If sheep are not well selected for breeding from, a tendency will be noted towards short wool in front and long, stringy wool behind. The ideal fleece should be of uniform staple throughout; but although this is an impossibility, an approximation thereto may be effected by careful selection of breeding rams, and, if possible, by the culling of ewes. The expense of wool sorting is considerable, and must be added to the depreciated value of the wool in poorly-bred fleeces, so that the superior financial return resulting from good breeding is most obvious.

The typical long-wool breeds of sheep by no means produce identical wools. The differences between the Lincoln and Leicester wools are not so great as the differences between sheep of the same breed fed differently—say, some on grass, &c., and some on root crops. But, on the other hand, there are certain characteristic breeds of long-wools worthy of special consideration. The Wensleydale sheep, for example, produces a wool of a wonderful lustrous, curly staple, and is a prime favourite among wool staplers. The Devon long-wool should also be specially considered. Thus, in the longwool lustre class are many varieties, each of which is worthy of special consideration from the wool point of view alone. Perhaps the tendency of the Wensleydale sheep to produce black lambs should here be referred to as a race characteristic. From observations made over a period of years—though not experimentally confirmed—it would appear that the Wensleydale black sheep is, in Mendelian language, a "recessive," whereas the Welsh mountain sheep, when black, appears to be "dominant." If it were a question of developing a black breed of sheep, this information would be most useful. For example, taking the Wensleydale black sheep as a basis, it would breed true black with black, and although when mated with white, would disappear in the first generation, it would reappear as black in the second generation, as the following table indicates.

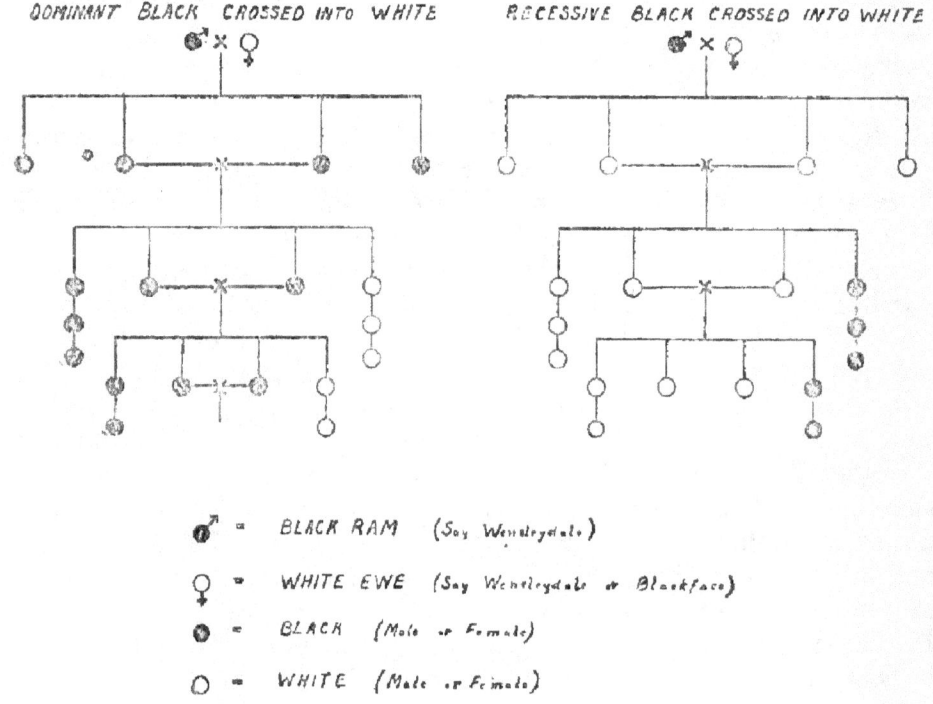

DOMINANT BLACK CROSSED INTO WHITE   RECESSIVE BLACK CROSSED INTO WHITE

♂ = BLACK RAM   (Say Wensleydale)

♀ = WHITE EWE   (Say Wensleydale or Blackface)

● = BLACK   (Male or Female)

○ = WHITE   (Male or Female)

So that from one black sheep only it is quite possible to develop a black flock by careful in-breeding with white.

The problem of the pure black sheep, however, is a very simple problem in comparison with the problem of grey or dark hairs in white wool. Certain breeds of sheep have a most marked tendency to produce these grey hairs, and as such wools cannot be used for pure whites, their value is markedly depreciated. The writer has seen staples of wool white for a certain length and then black. This seems to suggest that the pigmentation of the naturally-coloured wools is independent of their growth. If so, it is conceivable that means might be found of arresting the production of pigment, with a concomitant increase in the value of the wool as " pure white."

Reference has already been made to the effect of food on the wool. It is popularly supposed that if the Lincoln sheep be taken from South Lincolnshire into Norfolk—a distance of a few miles—it at once loses its lustre and markedly depreciates in value. Environment undoubtedly does affect wool to a considerable extent, and the above supposition may be true. It is to be regretted, however, that there are no authentic records of the effects of such change in environment. Obviously, the matter has a two-fold aspect— firstly, what will be the effect on the subsequent fleeces produced by an individual sheep under the changed conditions; secondly, what will be the effect on the fleeces produced by the progeny of such sheep. In other words, is the influence of environment direct or indirect ? There is here obviously room for some very interesting experiments.

The nature of the soil upon which the sheep is reared has another very important influence on the value of the wool—it may seriously affect the "yield." In the "Wool Year Book," for example, the yields of Lincolnshire wools are given as 80 to 82 per cent., but the writer has come across Lincoln wools turnip fed losing much more than this, even in its marketable-state. Of course, washing conditions may vary, but the farmer should see to it that his sheep are so washed that the full lustre of the wool is revealed —in the long run it will pay him to do this if his sheep are well bred on the lines here laid down.

Typical Short-wools and Short-wool Breeds of Sheep.—The Southdown is the typical short-wool British sheep. Closely allied to it are the Suffolk, Hampshire, and Oxford Downs, producing wool of a comparatively short staple—say 3 to 6 inches—fine of fibre in comparison with the long-wools and remarkably "frizzy." Perhaps the most valuable characteristic of Southdown wool is its fineness ; indeed, could it be bred slightly finer, its value would again be markedly increased. Reference has already been made to the possibility of rearing Merino sheep in the British Isles. It seems probable that the insufficiency of the return in mutton and wool has been the dominating influence rather than the impossibility of growing the wool in this country. Merino flocks have been kept for a considerable period, but again there are no records of the effect of our climate on (*a*) the individual sheep, and (*b*) on its progeny. The casual opinion of those who have had Merino sheep in hand in this country is that the wool does not deteriorate, but rather becomes finer. There is the question of food, however, and when it is realised that even in Australia there is a difference of opinion as to the effect of food on the fleece, it is obvious that no completely satisfactory assurance is likely to be obtained in this country, as the Merino sheep is never likely to be seriously taken in hand by the British farmer. The finest Down wool is not quite as fine as the coarsest Merino wool, while the coarser Down wools—Oxford Down, for example—are considerably coarser in fibre than the coarsest Merino. In view of the fact that fineness in fibre is one of the most valuable characteristics in these wools, the question arises as to whether shearing the lambs, say, at six or eight months' growth, might not yield a specially valuable type of material. This remark would apply not only to pure Down wools, but to varieties of the Down wools such as Oxford and Shropshire wools, in which the ultimate and natural growth being longer than in the case of the pure Downs, a longer lambs' wool is naturally yielded.

As these wools are carded, combed and spun into yarn on different lines from the long lustre wools, length is not an all-important factor, but reasonable length is certainly desirable. This must not be obtained, however, at the cost of fineness of fibre, and it is just as important in these fleeces as in the case of Lincoln, etc., that the wool should be as uniform as possible. The specially curly, frizzy characteristic of Down wools is of particular value in the manufacture of certain hosiery, etc., goods, and should most certainly be maintained and developed if possible.

The remarks made with reference to black hair, etc., in the long lustre wools are of equal or greater importance here, as certain Down breeds are specially noted for this defect. Careful elimination of black individuals and of all sheep showing any tendency to produce black hairs among the white should be the rule. There is also supposed to be a difference in the whiteness of Oxford Down as compared with Hampshire Down wool, which the writer has not had the opportunity of investigating; but it may be taken as a maxim that the whitest wool is most distinctly the best, and every endeavour should be made to select the whitest woolled rams for breeding purposes.

The conditions under which the Down breeds are reared no doubt affect considerably the wool. The grassy downs of the Southern counties are admirably adapted to this breed of sheep, or rather this breed of sheep is admirably adapted to this type of land. Curiously enough, however, flocks of Cheviot sheep are to be noted even in the South, so that the Down breeds undoubtedly have a competitor. The relative merits of the two breeds for grassy slopes such as are to be found in the Cheviots and in the Southern counties refer, perhaps, more particularly to the mutton, and the question is probably one of the carrying properties of the land : it is conceivable that three Cheviots might be more profitable than two Downs. The half-bred Down has reached Yorkshire, but does not yet appear to have made much progress further north.

Thus it would appear that there is a marked connection between the land and the sheep and the wool which may be most satisfactorily produced upon it. Heavy lands seem most suitable for heavy sheep producing long lustre wool, and the lighter and more chalky lands seem specially adapted to the requirements of the Down breeds. To select the breed most suitable for a particular environment is the great thing to attempt, and in doing this wool characteristics should be most carefully considered.

Variations from the two foregoing Types.—Three classes of variations from the foregoing types may be noted, viz., Half-breds, Cross-breds, and Special breeds. Half-breds are produced by the crossing of two distinct breeds. The value of such crossing will be evident when it is realised that in Yorkshire, for example, there are large tracts of land which will not support a large-bodied sheep, such as the Wensleydale or Leicester, but upon which the smaller and more active Blackface will thrive. Blackface ewes are, therefore, run upon these tracts of land, but for the richer land in the bottom of the valley or for sending down to the plains for fattening up for the market a cross-bred sheep is bred, the Blackface ewes being mated with a Wensleydale ram, and thus a more valuable lamb and sheep produced than if Blackface were mated with Blackface. Further north, a similar crossing is practised with the Cheviot and the Border Leicester. This latter breed no doubt gives valuable mutton and wool characteristics to the half-bred sheep produced ; in fact, this cross is that from which the famous "North Wool" is produced which finds such a ready sale in the Bradford market.

Crossing undoubtedly upsets the stability of a breed, and consequently there is, perhaps, a reasonable tendency to be very conservative in crossing. If, however, there were some means of advising the farmers on the results of their crosses, it is quite possible that even more useful crosses than those at present prevailing would be discovered. It is possible that the conjoint action of the Agricultural and Textile Industries Departments of Leeds University in this matter might lead to most useful results. Unless the farmers can readily obtain reliable advice as to the mutton and wool values of suggested crosses, it is obvious that few new crosses will be attempted—the farmer will prefer to work safely upon the old lines.

The greatest detriment to crossing lies in the very varied wools produced. Sometimes the progeny of a cross revert to one or other parent, but more frequently a varied, and possibly, irregular fleece is produced. Thus, save in one or two special cases, the farmer cannot hope to make a name for himself as a wool producer if he indulges in cross-breeding. If, however, he cross-breeds with the idea of working back again to his pure breed, with a certain robustness added, there may be an advantage in crossing, even from the wool point of view. The method pursued in effecting such crossing in Australia is well illustrated in the following diagram, illustrating the production of " come-back " wool.

The crossbred wools of New Zealand are obviously the result of the search for a superior mutton sheep, and the same remark applies to crossbred wools, such as there are in this country. Thus in the case of the Blackface-Wensleydale cross a three-quarter bred sheep is often produced by mating the first cross with one or other of the original types. Such cross-breeding, however, is very liable to upset the wool characteristics of the animal, and although useful from the mutton point of view, is not to be commended so far as the wool is concerned. Of course, the wool thus produced, if compared with the Blackface wool, is better ; but in comparison with, say, the Wensleydale wool, is markedly inferior and more irregular. After all, the problem seems to be one of breeding a sheep to give the utmost return from the land in both mutton and wool.

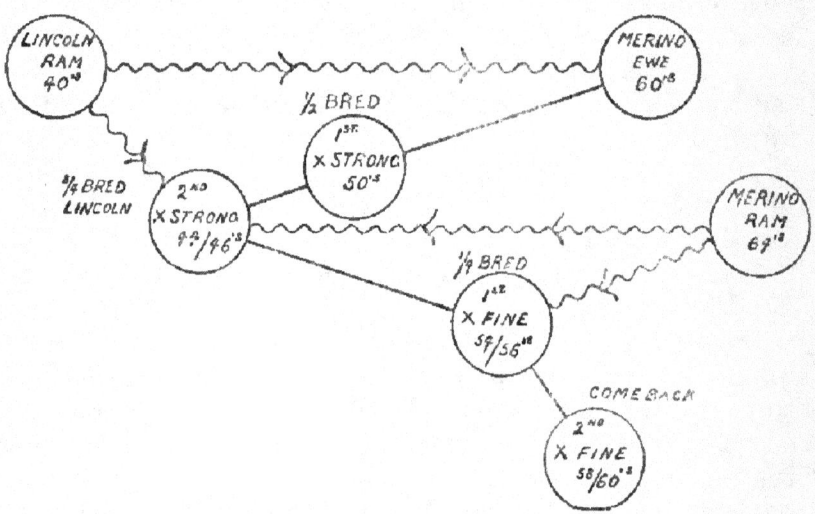

DIAGRAM ILLUSTRATING THE PRODUCTION OF AUSTRALIAN "COME-BACK" WOOL.

In some few cases, special breeds of sheep are cultivated with a particular object in view. Thus, the Shetland sheep supplies the Shetlanders with the beautifully long, silky, fine wool from which they manufacture the renowned " Shetland shawls." The Cheviot sheep may also be cited as a sheep whose wool has given rise to a special industry, viz., Cheviot tweeds, while the Welsh mountain sheep has produced a wool which, from its non-shrinking properties has become famous in the form of Welsh flannels. Thus again it seems most obvious that if our farmers were less conservative and acted under good advice, much benefit, both to themselves and the wool manufacturer, might accrue. For example, pure Shetland wool may be worth 5s. per pound for making into Shetland shawls, but if not adapted to its special purpose it is probably worth less than 1s. per pound. That there is so little connection between the wool grower and the wool manufacturer is lamentable, and it is hoped that the Wool Exhibit at the Yorkshire Show, held in Bradford last June, has stimulated the interest of all concerned, and prove the value of further co-operation in the interests of all parties concerned.

The following table, from " The Sheep Breeders' Directory," showing the views of the breeders of New Zealand as to the relative values of the different English sheep for crossing, may here be studied with advantage.

The Breed of Sheep.	Apti- tude to fatten.	Hardness and Soundness.		Lamb- ing and Inc'se.	Form and Shapes.	The Mutton.		
		Con- stitu- tion.	On feet.			Appear- ance.	Qual- ity.	Wgt.
Maximum Points ..	7	8	5	6	12	3	7	6
The Lincoln ..	6	4	3	4	11	2	5	6
The Border Leicester ..	7	6	4	5	12	2	6	4
The English Leicester ..	7	5	3	5	10	1	4	3
The Romney Marsh ..	5	7	5	6	11	2	5	4
The Shropshire Down ..	6	6	3	6	12	3	7	3
The Southdown ..	6	5	3	5	12	3	7	2
The Hampshire Down ..	6	5	3	5	11	2	6	6

CHAPTER XXVII.

IMPROVING WOOL AND MARKETING THE FLEECE.

Improvements in (A) *Mutton,* (B) *Wool, and* (C) *Mutton and Wool.*— From the foregoing particulars it will be gathered that while the mutton factor will be most frequently dominant, very occasionally the wool factor attains to first position, while frequently both wool and mutton may be developed together to marked advantage. Perhaps the broadest view, even if strange, is nevertheless the most illuminative way of looking at the problem—and that is to regard the sheep as an instrument, a machine, for turning grass into flesh, hide and wool. Then it is evident that the product from any given pasturage and sheep will depend upon (*a*) the nature of the pasturage, and (*b*) the type of machine—or rather breed of sheep—employed in the conversion. We should not expect the heather moorlands of Yorkshire to produce the same mutton and wool as the Downs of the Southern counties, or the rich root-crop lands of East Yorkshire and Lincolnshire. But when we come to inquire closely into this matter, we find that we cannot well dissociate the pasturage and the breed of sheep: the interdependence one upon the other is so marked that any independent differentiation seems hopeless. This is not altogether so, however. Barren lands will only carry small sheep, but the barren lands of Australia carry the small Merino sheep with its fine wool, while the comparatively barren Yorkshire moorlands carry the small Blackface sheep with its long, coarse wool. Again, the richer lands may carry the large Lincoln sheep with its long lustre wool, or the large Downs sheep with its short, crisp wool. In other words, it seems as though breed of sheep was the commanding factor, and that whatever the type of land may be, it is possible to obtain a breed of sheep which will produce both good mutton and a useful wool. One of the most striking instances of this has recently come to the writer's notice in the flocks of Mr. Elwes. The Herdwick sheep produces the roughest of English wools, suitable only for the low carpet trade. This sheep, crossed with the Shetland, produces a most wonderfully improved wool, with little or no detriment to the mutton. The difference between the pure Herdwick wool and the Shetland-Herdwick is so great that it is impossible to satisfactorily describe it in words ; the value of the wool is at least doubled, or, for special purposes, quadrupled.

An interesting attempt, already referred to, to breed a large carcase sheep with a heavy fleece of fine wool has been made on the Cambridge Agricultural Farm, two Merino rams having been mated with Shropshire ewes, the idea being at the second generation (F2) to select the sheep with the large Shropshire body and the fine Merino wool. The Mendelian basis for this experiment lies in the reshuffling of characters which occurs in cross-breeding followed by the artificial selection of the most favourable variety. Unfortunately, the characters of sheep are so many and so varied that it seems impossible to isolate any two or three desirable characters, as in the case of certain plants and animals with which Professors Bateson and Punnett have so successfully worked ; so that the characteristics of mutton, wool, &c., are so variously defined in the progeny of such crosses as the Merino-Shropshire that no satisfactory selection seems possible. In the first generation of this cross it was noticeable that the finest-woolled ram was the smallest bodied and most delicate, and consequently was useless to breed from with both mutton and wool in view, while further crosses seem to produce no further segregation of either the wool or mutton characteristics, probably due in the first place to the Merino and Shropshire coming in part from the same stock, and partly due, as just explained, to the multitudinous and mixed characters present in even the most characteristic breeds of sheep. Professor Wood's experiments in crossing the hornless blackface Suffolk with the horned pinkface Dorset, however, suggest that Mendelian principles may yet be of use to the sheep-breeder, and although too much must evidently not be expected from the application of these principles, it would obviously be foolish to throw the theory overboard altogether. With practical men, the wish is too often the father to the thought and to the action ; prejudice is rampant, and too often they are unaware of their innate desire to confound every novel suggestion, every out-of-the-common possibility. It is more than probable that, with more hearty co-operation on the part of sheep-breeders, much light might be thrown upon the very important influences of race and environment in heredity.

The Classification of Wools.—Broadly speaking, wools are valuable as possessing :—

(a) Lustre and length, or
(b) Fineness and waviness.

The long lustre wools may be spun and woven into certain characteristic fabrics, such as lustre cloths and bright serges. The so-called "Alpacas" are not infrequently made from long lustre wool.

On the other hand, the fine, wavy Down wools may be spun into soft handling yarns which may then be knitted into hosiery goods or woven into soft, full-handling tweeds. Obviously, then, the sheep-breeder should endeavour to bring his flock distinctly under one or other of the above categories if the greatest wool value is to be obtained. Thus, one farmer may find that his land will carry Lincoln sheep. It would also carry, say, Oxford Down sheep. But if he were to cross these two breeds, the resultant wool would be neither a good long lustre nor a good short, wavy wool; it would fall into a nondescript class between the two and probably be much less valuable than either. Of course, according to fashion, the relative values of long and short wools may vary, but broadly speaking the foregoing remarks hold true. Thus, although it may be true that by far the greater weight of wool grown in the British Isles is neither one nor the other of the two foregoing types, but something in between, it is nevertheless useful to bear the two types in mind and to approach as near as possible to one or other of the types as opportunity offers. Shropshire wools, for example, have Down characteristics combined with length in the full-grown fleeces. Thus, length may be and frequently is an actual obstacle to its treatment as a soft tweed or hosiery wool, and its Down characteristic markedly decreases its value as a long lustre wool. What more natural, then, that Shropshire lambs should be sheared and a shorter Down wool of greater value thus produced. It is not sufficiently recognised that under certain circumstances wool may be too long. Thus, combers may sometimes be instructed to comb out not the short fibres (or "noil" as it is termed) but the long fibres, thus producing a short type average top.

Having decided upon the type of wool to breed to, every endeavour should be made to obtain uniform fleeces. In certain crosses, and possibly in all fleece (subsequent to second year) wool, there is a marked tendency for the hinder half of the fleece to be coarser than the front half. This should be corrected so far as may be, and most strenuous endeavours made to produce a uni-fleece.

The Marketing of Wools.—In the long run, the large sheep-breeder will find it to his advantage to classify his fleeces and to face the wool buyer with "well got-up wool." He must not expect to reap any advantage by doing this for one year only; he must rather hope to establish a name by persisting in the right getting-up of his wool year by year. The wool must be well washed on the sheep's back, cleanly sheared without double-cuts, free from straw and vegetable bits, tied up with itself and not with stack-band or other objectionable vegetable fibred stuff, dirt and foreign matter removed, and a normal amount of moisture only present. Black and grey fleeces must be kept separated from the white.

If the farmer will pay attention to these points, then it may further pay him to co-operate with others in the public auction of his fleeces ; but if he is not prepared to take pains in the manner suggested, then he had better let well alone, and resign himself into the hands of the wool man who travels his country buying up wools for the Bradford wool stapler ; the chances are he will be as well treated as he deserves.

The importance of the satisfactory " get-up " of the fleeces in the eyes of wool buyers is well illustrated by the following extract from the report by the representative of *The Yorkshire Observer*, on the Kettering Wool Sale on June 11th :—

" The character and quality of the half-breds were generally fairly satisfactory, and the grass-grown lots generally were clean and bright, and well prepared for the market. It was, however, unsatisfactory to find some large lots of wool in a greasy and un-washed condition. These clips had not been at all well secured, and, in fact, they constituted almost a gambling transaction to buyers. Being grown in a district where ironstone abounds almost on the surface, the unwashed wools were impregnated with red pigments and weighed almost as heavy as lead. It was no wonder that unwashed half-bred and Down fleeces were on the average about 9s. 6d. per tod of 28 lb. below the rates which would have been secured had they been washed. The only and poor consolation the sellers had was that buyers were made to pay not for wool only but for wool plus heavy grease and still heavier iron-stone sand impregnated in the fleeces.

" Side by side with this unsatisfactory feature there were lots which were highly meritorious in the way in which they had been secured and brought to market. Even the wool grown on the " red " land when well washed looked and handled well, while the best washed grass-grown wools were all that could be desired, both as regards quality and condition."

Wool Qualities.—In conclusion, a word or two on wool qualities may be useful. The spinner constantly refers to his wool by quality numbers, an idea of the meaning of which may be useful to the wool-grower. The quality number is supposed to indicate the count of yarn or fineness to which the material will spin. Thus :—

a 28's quality wool should spin to $560^* \times 28 = 15,680$ yds. per lb.
a 40's ,, ,, ,, ,, $560 \times 40 = 22,400$,, ,,
a 60's ,, ,, ,, ,, $560 \times 60 = 33,600$,, ,,
a 90's ,, ,, ,, ,, $560 \times 90 = 50,400$,, ,,

The quality number, however, is not entirely reliable from this point of view. It is very questionable, for example, whether a 28's britch, as it is termed, will spin to 28's count of yarn (that is 1 lb. of wool drawn out to 15,680 yards). On the other hand, it is probable that a 70's quality (merino wool) will spin to a 80's

---

* 560 yards in the worsted hank or unit of measurement.

or 90's count of yarn (that is, 1 lb. drawn or spun out to 44,800 or 50,400 yards per lb.). In other words, although the spinning capacity idea was the fundamental basis, the quality numbers are now more or less artificial standards, fully recognised in the trade, but somewhat difficult to get hold of and to be certain about on the part of the uninitiated.

From these particulars it will be evident that the various breeds of sheep will produce wools with various quality numbers. Thus —

Lincoln Hogg fleeces range from		40's to 28's	quality.
„ Wether „	„	„	36's to 28's	„
Kent fleeces	„	„	46's to 32's	„
Cheviot „	„	„	50's to 36's	„
Southdown fleeces	„	„	56's to 32's	„

There is usually some relationship between quality number, length and fineness. For example, 28's britch Lincoln is long and coarse, 32's Lincoln is comparable with 32's britch Kent, and 50's Cheviot and 50's Oxford Down would also be comparable. But there is no comparison in length between 32's Lincoln and 32's Southdown ; here coarseness of fibre is the deciding factor.

These brief particulars will give the wool-grower an idea as to what quality in wool really betokens and also how to estimate the quality in his own wools. Should he wish to go more deeply into the matter, Messrs. Cassell's publication on " Wool Carding and Combing " will just give him the information he needs. In view of the interest taken in this matter by Colonial sheep-breeders, and their constant attendance in the Textile Industries Department of Leeds University and Bradford Technical College, English sheep-breeders might well consider whether it would not pay them to take short courses at one or other of these institutions. Possibly a course in conjunction with the Agricultural Department of Leeds University could be arranged with advantage.

CHAPTER XXVIII.

DISEASES OF SHEEP.

Many of the diseases of sheep are common to other ruminants but in the treatment of their ailments sheep must needs, as a rule, be dealt with in a wholesale manner; the individual being of less value than the ox and not under the same discipline or management as the cow accustomed to handling. Readers, with experience of flocks, will readily call to mind the trough for foot-dressing and the drive through a narrow gateway in which remedies are placed with which the feet may receive a rough dressing as the animals tread it. The conditions under which a flock is kept apply to all the animals composing it, and they have been exposed to the like chances of disease, so that medicaments are often mixed with food in the trough from which all partake in something like the same proportions; a rough method truly, but large flocks cannot well be dealt with otherwise. The small holder with a few animals can give the individual attention which is to be preferred, and the ewe flock is placed under special conditions for lambing and men's services set apart. In the following pages, we shall consider the common diseases with the general methods employed upon large numbers, as well as the treatment suited to individuals under complete control.

Signs of Disease.—The gregarious habit impels a sheep only slightly ailing to keep up with the movements of the flock, but the watchful shepherd will notice, if any of them keep their heads up while the rest are grazing, that something may be assumed to be amiss when any departure from habit is observed. A sheep more seriously ill will be left behind, or intentionally leave his fellows. He should be caught and examined. The ailing one may show other signs such as arching of the back, extended neck and lowered head and drooping ears, or be found prone and making no effort to rise. In certain brain disturbances they appear to be dazed and lost to the world. Indifference to surroundings, however shown, is a sign of illness. Fits in which the sheep spins round and falls over, or holds the head to one side for some time and presently sinks to the ground suggest to the flockmaster the presence of brain pressure from hydatids, as does any want of co-ordination of the muscles or stumbling which will not be mistaken for lameness any more than the irregular movements of an intoxicated person. While speaking of individual sheep unable to rise, it would seem superfluous to say that fat animals of the large breeds, with heavy fleeces, sometimes get on their backs and perish in useless struggles to right themselves, for this is known to all practical sheep men; but we have seen farm hands so little interested in what does not happen to be their work, that they would pass an animal in this helpless condition without taking the momentary trouble of turning the unfortunate creature; an act of charity that any passing stranger should perform.

The signs already mentioned as affecting individuals may be wanting, and yet a flock be in serious danger of disease, as when first invaded by the liver parasite called fluke. They will not all suffer in the same degree, but the trained eye will observe a failure to improve at the rate warranted by the keep or feeding, and here is the great value of a good shepherd who will immediately call attention to troubles that may be dealt with promptly, but soon pass beyond remedy if not early detected. The suspicions of the shepherd should lead to investigation, and it often proves the most economical method to slaughter and make a *post-mortem* examination of one or two of the first to fail with what may be a disease afflicting the whole flock, more or less. Any signs of constipation or of the opposite condition of diarrhœa must not be passed over as of no consequence. A scrubby mountain side in dry weather and the consumption of much woody fibre and in-nutritious stuff may be the cause of the former, and a sudden transference from dry food to cabbage or turnips or a lush pasture may account for laxity of the bowels on the other hand, but scour may be a sign of tapeworm, or of those myriad strongles which have decimated many flocks. Stained quarters should be regarded with suspicion, and the person in charge of sheep should make a habit of observing their excrement, breaking up hard fæces and looking closely into their composition with an eye for segments of tapeworm and the fine, thread-like creatures alluded to as strongles. Digestive disturbances are often shown by blowing out of the flank or tympanites or the reverse condition of hollow flank, indicating that want of moderate distension that is essential to good digestion in all ruminants. Discharges from the nose, or watering of the eyes have the same significance in sheep as in other animals, and catarrh so recognised should lead to changed conditions or medical treatment. Frequent posturing to urinate without the required relief, or discolouration of the fluid, should lead to examination and, perhaps, to different management of the flock. Skin troubles are recognised by rubbing against fixed objects, and the coming out of the fleece is a sign that the mischief is more or less advanced. Irritation of the skin from scab mites or by being struck with the fly is often shown by frequent shaking of the tail. In the chapters on diseases, the symptoms taken together will be more fully described; those we have mentioned are such as the flock owner and his man should be always on the look out for, and by early attention prevent further trouble.

CHAPTER XXIX.

DISEASES OF THE BLOOD.

Estimates of the blood in sheep lead us to suppose that the proportion to weight of the carcass is as one to twenty-three. The composition, as regards red blood corpuscles, is capable of greater variations than in perhaps any other of the domesticated animals. A condition of plethora or of full bloodedness rendering these animals specially liable to such diseases as anthrax or strike, and to sudden stases resembling apoplexy when they are over-fed with stimulating foods, with a view to showing in fat stock shows or for early slaughter. There is reason to believe that a greater number of cases of anthrax occur among sheep than has been generally supposed, as deaths among the full-blooded are attributed to the plethoric condition just alluded to, but this state has only prepared the sheep for the invasion of the specific bacillus. The condition known as anæmia, in which the number of red blood corpuscles sinks to the lowest known among domesticated animals is the other extreme, and it is quite remarkable how long sheep can keep alive when starvation or disease has robbed them of these important red cells.

The heart-beats, as measured by the pulse, which can be taken under the tail (caudal arteries) or in the thigh (femoral) is variously estimated at from seventy to eighty per minute in adults, and eighty-five to ninety-five in lambs. The temperature is commonly said to be 101° to 104° Fahrenheit. The sheep doctor will not attach so very much importance to these figures because the timid nature of the sheep accelerates the heart's action when restrained for examination, and the same nervous influences elevate the temperature. Unless the pulse is a mere flicker on the one hand (as in anæmia) or full and bounding as in brain troubles and plethora on the other, the examiner will take notice rather of the signs mentioned in the previous chapter, and of the colour of the visible membranes and the fullness or otherwise of the vessels of the eye when the lids are forced open. A temperature much over 104° Fahrenheit, will, however, be considered serious.

SIMPLE FEVER.

A febrile condition in which there is a loss of spirit and of appetite and absence of rumination, dry muzzle and hanging head, quick pulse, constipation, and general ill-doing has been generally recognised by sheep men, but so far no specific organism has been credited with its causation. Persistent fever of this kind leads in some instances to jaundice, and then the yellowness of the skin and the membranes of the eye and nostrils declare the nature of the organ most affected. The urine is also bile-stained.

Treatment.—Two or three small doses of salts, with gentian and nitre. Half an ounce, two drams, half a dram respectively for each sheep, in the day's ration. If the numbers permit of handling a solution of quinine, in dilute sulphuric acid, and in water, may be given as a drench. The dose should contain five to eight grains of the drug. A dry upland and generous trough feeding enable the animals to pick up after the febrile symptoms have passed away.

ANÆMIA.

It takes much longer to supply the required building materials than to deplete the plethoric animal. The full blooded may be purged and blood tension lowered by reduced rations, but the anæmic must not only be steadily fed with nutritious food, but supplied with those substances best known to favour the manufacture of red blood corpuscles. Iron and salt supply these, and should be given to sheep showing the symptoms of pallor of the membranes, short-windedness and dropsy under the jaw or chog. Anæmia is of two kinds : that dependant on debility from ill-doing or want of suitable food and unfavourable climatic conditions, and a so-called

PERNICIOUS ANÆMIA.

This is believed to be due to some microscopic parasites of the same nature that causes fowls to go light. Whatever the cause, the result is the wasting away of the red blood cells and their almost entire disappearance, with consequent debility affecting the vital organs. It is usually, but not always, accompanied by wasting of the fat reserves.

Symptoms.—In the flock, the first attractive symptom is roughness of the fleece, weak behaviour in moving, or reluctance to seek food, and pick and choose among the pasture plants, which is a feature of the vigorous, already satisfied, but not disposed to ruminate for awhile. Wool coming out, or held loosely to the skin, is a symptom common to many diseases, and should lead to further and closer examination. If pernicious anæmia is the cause, the sheep will then be found to have pale membranes, a cool and damp tongue, perhaps palpitation of the heart when first caught, and subsequent heart failure, to the point of fainting when held awhile. Dropsical swelling between the lower jaws, known as poked or chockered in parts of England. This symptom also belongs to the debilitated from fluke and other parasitic invasions, and to arrive at a correct diagnosis, all the circumstances must be reviewed. It may be that the sheep have had too many turnips, grown in a sunless season, or that one kind of food may have been given too long, or, as previously suggested, some piroplasm is at work.

Treatment.—Change the food and change the fold or place. Feed generously on a mixed diet. Cake and corn should be among the articles given. For medicine, table salt (chloride of sodium), sulphate of iron, and ground gentian root. A dose approximating to thirty grains of the finely-powdered sulphate of iron, forty grains of salt, and ninety grains of gentian, should be estimated for as each sheep's portion daily. It should be well mixed with cake dust or other food in a fine state of division, as the administration of drugs in the trough is, at best, uncertain, and we wish to avoid overdosing a few and depriving others of their share. An insufficient number of troughs contributes to this trouble. These drugs may give place, in a week or ten days, to two drams each of powdered cinchona bark, and fenugraec, continuing the salt. By such treatment many will be saved, but pernicious anæmia, unrecognised in the early stage often proves fatal, and the reader is warned in this, as in other sheep diseases, of the necessity of constant watchfulness over the flock.

RED WATER.

Sheep men do not always mean the same thing as cattle men when they use the term "red water." They refer to a red accumulation of dropsical fluid in the abdomen, which is only discovered after death. There are several causes probably. Microscopic examination of the fluid has shown the presence of organisms of the kind that cause heart-water and other serious effusions, but it is not known how they enter.

The symptoms of this malady are not very easily distinguished from other forms of dropsy, and the nature of the disease is not known until a dead animal is opened.

Treatment.—Removal to fresh ground, mineral and vegetable tonics as advised for anæmia, which see.

RED WATER.

The passing of red-coloured urine as in cattle with the disease known by the above name. This discolouration is the diagnostic symptom.

Treatment.—A purgative such as a dram or two of aloes, half an ounce of bicarbonate of potash, and three to six ounces of salts. Purgative doses of salt (chloride of sodium) which in many districts are favoured for cattle with this malady, should *not* be given to sheep, as they are specially sensitive to salt in anything but

P

alterative doses, such as we have advised in connection with anæmia or wasting diseases. Gruelling of individuals drawn out of the flock (large numbers are seldom attacked), and tonics of one to three drams of Peruvian bark and gentian, may be given at one end of the day, and thirty or forty drops of spirit of turpentine, in a tablespoonful of linseed oil, at the other. Recovery is the rule when placed under favourable conditions.

ANTHRAX.

The first intimation of the disease is, too commonly, a dead sheep, much blown up, and with a blood-stained discharge escaping from the natural orifices of the body. The cause is definitely known to be the anthrax bacillus. It may enter by a variety of portals, but probably most often by the mouth, in the food.

Symptoms.—The illness is of such short duration that signs are seldom observed. Standing apart, with flanks filled up like a blown animal on wet clover, and the appearance of being struck for death ; a fatal blow has been struck at the animal's vitality, and a person who has not seen it before would form the opinion that the animal must die. A state of collapse, yet often fearing to lie down, on account of the tympany or hoven referred to above.

Treatment.—Anthrax is a notifiable disease, and treatment is, therefore, out of the question. It is doubtful if it ever succeeded when it was permissible. Preventive measures are generally believed to be worth a trial, and this emphasizes the importance of correct diagnosis when a sheep is found dead or fatally struck. They consist in lowering the blood tension, as it has been previously stated in remarks on the blood that a full and rich-blooded state of the sheep predisposes to anthrax. A purgative then should be administered first, the ration reduced, or a change made to comparatively bare pasture. The hyposulphite of soda has a reputation as a preventive, and may be given in doses of half to one ounce for each sheep, in the powdered form, mixed with the trough feed daily.

QUARTER-ILL, SYMPTOMATIC ANTHRAX, STRIKE.

By these and other names a somewhat similar disease to anthrax is known. It is not caused by the same bacillus. Is not so rapidly fatal or disposed to spread; more often affecting an individual from time to time than occurring as an outbreak. No treatment is of use when the struck animal is discovered, but the same measures above advised for the prevention of anthrax can be used against the strike. There is much confusion about these blood diseases among sheep, and it may yet prove that there is some connection between those already considered and the two last mentioned.

BRAXY, SHEEP SICKNESS, STRIKE.

The last sentence of the above paragraph is made more forcible by perusal of any veterinary literature connected with sheep and braxy. All sorts of theories have been put forward only to be refuted, and many cases of anthrax must have been put down to braxy. All sorts of causes have been suggested, including a charge against the Gulf Stream for coming too near to that part of Scotland where the malady most frequently occurs. Its course has not materially altered since. That it is due to a specific bacillus of braxy seems now to be established. It was thought that Professor Hamilton had discovered a satisfactory means of prevention, but this has not stood the test of time.

Treatment.—This must be of a preventive kind, as the malady is almost invariably fatal when once developed. A very crude method of immunisation at present holds the field. It consists in putting a pig on an infected pasture, and mixing a pint of its dung with fifteen pints of milk, and straining the fluid. Of this elegant mixture, a wineglassful is administered to each sheep after a twelve hours' fast, about the third week in September. On the principle of attenuation, by passing the disease through another species, this practice is commended—in the absence of a better.

CHAPTER XXX.

DISEASES OF THE BREATHING ORGANS.

CATARRH.

Sheep suffer from colds or common catarrh like other animals, but it is rarely caused by exposure to extreme cold. Rather should we look for it as the result of travel, the irritation of dust, and the breathing of bad air in confinement.

The symptoms are running of the eyes and nose, at first of a watery fluid, and afterwards a thick or gummy and adherent matter around the nostrils.

Treatment.—Restoration to pure air and good conditions is all that is required, but debility resulting from prolonged catarrh will need tonics such as have been already suggested in connection with anæmia.

NASAL GLEET OR OZOENA.

A persistent discharge of variable consistence may be mistaken for catarrh when it is caused by the irritation of parasites. The sheep bot or grub lodges in the chambers, high up, as does another parasite known as a pentastome, and while it remains no external measures will be of use.

Treatment.—Sometimes the sneezing induced by a pinch of snuff will dislodge the enemy, but more often a safe refuge has been chosen, and we have to wait for the time of year when these creatures voluntarily quit, in order to undergo another stage of development. Preventive measures, then, must be relied upon. The bot fly must be prevented from ovipositing in the nostril, by causing the sheep to smear the parts with such things as fish oil and tar when seeking food through auger holes made in a plank to cover the provender in a trough. When sheep bunch together, with heads turned inward in hot weather, it is to avoid the attentions of the female fly. Syringing has been tried and inhalations of sulphur fumes, but success has not attended these methods. The numbers are kept down by the smearing methods above advised.

PLEURISY.

This is a more frequent disease in sheep than is realised by flockmasters, because recovery without treatment is the rule. The evidence remains, and the butcher finds the adhesion in the chest.

Symptoms.—A short, hard cough and rapid, shallow breathing, standing for the most part, and indifferent to food during the first stage. When the pleural membranes pour out fluid the more urgent symptoms pass away. If the chest does not fill up with " water " the animal recovers. If death results, the presence of much fluid is diagnostic of the malady.

Treatment.—Rubbing in of one part turpentine spirit, to four parts of seed oil, on the sides of the chest. Good nursing or hygienic conditions are of the greatest importance, such as protection from the weather, and pure air, with gruel and other sloppy foods that can be easily taken without keeping the head down or coughing by reason of dust or pollen grains.

PNEUMONIA OR INFLAMMATION OF LUNGS.

This is not so frequent a trouble with sheep as pleurisy and bronchitis.

Symptoms.—Shallow and rapid breathing, high temperature, loss of appetite, and a suppressed, painful cough, standing about with extended neck and anxious expression.

Treatment.—The same as for pleurisy, which see.

BRONCHITIS.

In ewe flocks or other sheep that are not quite young, a form of bronchitis of the more or less chronic kind is seen, and there is not much in the way of treatment that can be recommended, but when it takes the form in younger animals of acute attacks, a strong liniment should be rubbed down the course of the pipe, and behind the shoulders. It may be advisable in some cases to clip the wool in order to apply it. After this special care must be taken to avoid fresh chills. Medication takes the form of administering such drugs as half-dram doses of compound ipecacuanha powder, daily.

VERMINOUS BRONCHITIS OR HUSK.

This is the form of bronchitis we have most to fear, and among lambs or young sheep of the previous year. It is caused by the presence of worms in the smaller bronchial tubes. It sometimes causes death, and in any case is a serious loss to the flockmaster by throwing back the young animals, arresting their growth and development.

Symptoms.—A frequent husking or prolonged soft cough, foaming at the mouth and nostrils, and more or less distress and interruption while feeding.

Treatment.—Keeping up the general health and strength enables the lamb to make a good fight against the parasites, and for this purpose extra good food should be provided, and tonics, such as iron quinine and gentian mixed with the trough feed in small but regular doses. The administration of turpentine in oil undoubtedly hinders the development of the worms, as the volatile drug is exhaled to a small extent. Doses of forty to sixty drops or more in a wineglass of linseed oil, may be given at intervals of a few days. No treatment hitherto adopted is so successful as that of inter-tracheal injection of such agents as carbolic acid, turpentine and chloroform. Weak solutions of perchloride of mercury are also injected by the hypodermic syringe into direct contact with the worms. A veterinary surgeon is usually employed for this work, and it should not be undertaken without some instruction from an expert.

CHAPTER XXXI.

DISEASES OF THE DIGESTIVE SYSTEM.

The anatomical arrangement of the sheep is, in the main, that of the ox, but there are certain differences and peculiarities, such as the cleft which divides the lip and enables the ovine species to bite close and thrive on a pasture that would not support cattle. This close cropping causes lateral growth, and so benefits the land, beyond the manurial elements deposited ; and these are of much value. The sharp cutting edges of the teeth which help so much in the cutting of fibrous foods and the scooping out of "roots" later on in life, lead to dental troubles seldom seen in cattle. "Broken mouthed " is a common description of the old ewe. The cutting teeth are sometimes broken when dealing with frozen turnips and other resisting substances, and lodgments and irritation of the gums follow. Besides such injuries, the teeth wear and come out by the insinuation of gravel and sand, and in those animals where perhaps less than half the number are left, it may be better to extract the remaining more or less loose incisors and let the gum harden off, as it will, and serve the purpose fairly well.

APTHA OR THRUSH.

A simple eruptive disease affecting the mouth and lips of lambs, and in some seasons of adult sheep, has long been recognised. Crops of vesicles following one another and breaking up and leaving slight ulcerations of the mouth membranes. The thin portion of the tongue at its sides is most often affected.

Symptoms.—Sore mouth and difficulty in feeding, dribbling of stained saliva, slightly tinged with blood at times, redness of the mouth and whites of the eyes, and in some individuals considerable fever and loss of appetite, which in part may be due to soreness of the mouth and pain endured in trying to browse. Such animals wander about and have a dejected appearance, without the signs of real illness or being smitten in the way referred to in connection with blood diseases, which see. When caught up, the breath is found to be offensive in odour, and the subjects more or less wasted in flesh.

Treatment.—This form of simple aptha readily yields to medication ; all that is necessary is to give two or three small doses of salts, with bicarbonate of potash, and dress the membranes with borax and honey, in the proportion of one of the former to seven of the latter. If healing does not rapidly take place, a solution of alum may be substituted, such as a quarter ounce dissolved in a pint of rain water, and mopped on to the sores daily.

MALIGNANT APTHA.

This is a much more serious matter. It is not only very infectious but rapidly lowers the vital powers and in many cases proves fatal. Suckers convey it from the mouth to the teats and udders of their mothers, and this leads to complications and sometimes garget.

Symptoms.—Those described in simple aptha, but greatly exaggerated. Pustules take the place of simple vesicles, and badly ulcerated surfaces are left. The trouble extends to, if it does not actually originate in the stomach (*stomatitis*), and the whole body is sick.

Treatment —Immediate isolation of the sick and of all suspects.

Disinfection.—Dressing of the sores with a weak carbolic lotion, such as a dram of carbolic acid, two drams of glycerine, and a quarter-pint of water; an adult may be given half a dram of chlorate of potash, with a dram of hyposulphite of soda twice a day, dissolved in water ; lambs, in proportion to their size and age. Sulphate of magnesia in half to two-ounce doses at intervals of a few days may be advisable. The food should be moist and nutritious and easily mouthed and swallowed. Cooked linseed and bran and sharps and the like being suitable for sustaining the patients, but it must never be forgotten that ruminants cannot get their cud without some bulky food. Scalded hay will serve this purpose, without hurting the tender membrane.

It has been said in another place, that a good deal of confusion exists about sheep diseases, and this unfortunately applies to these eruptive diseases of the mouth. The reader will not insist upon correct classification—a thing at all times difficult—but may be helped in his diagnosis if we here refer to another malady which is often mistaken for the simple and the pustular or stomatitis above mentioned. It is called by many names and among them

CONTAGIOUS PUSTULAR DERMATITIS, LIP AND LEG DISEASE.

It is one of those maladies which the bacteriologist has thrown light upon that enables us to apply his knowledge to its cure.

Symptoms.—Fever to a varying extent. The lips and face, and very likely the legs and feet show an eruption of a painful nature, which is disposed to spread rather than to heal. Its variable situation and the order in which it attacks the various structures have made it known in some districts as " mouth and foot," " hair and hoof," and was at one time described as " carbuncle of the coronary band." We know now that the bacillus necrosis is the cause, and that the suffering mouth applied to the leg or pastern will convey to those parts : or the irritation of the pastern, causing the sheep to lick or rub with his muzzle, will be the means of infecting the mouth. The flockmaster generally fails to recognise it at first, but becomes aware of something worse than ordinary aptha or the stomatitis before mentioned, because the swelling and grey incrustation is followed by loss of substance. An ulcer is a loss of substance in a more or less circumscribed area, but this necrosis is a general destruction that keeps spreading. It sometimes attacks the membrane of the cheek and destroys a portion, so that food accumulates and brings about a bulge which attracts notice.

Treatment.—The destruction of the causal organisms is essential to success. This needs care. The application of a piece of lint dipped in a 10 per cent. solution of chloride of zinc has given good results in skilled hands. It should only be used once, after which an ointment for external lesions may be used, such as carbolic acid five parts, flowers of sulphur ten parts, goose grease or vaseline 100 parts. Mouth damages should be treated with one part alum, three parts glycerine ; iron, salt and gentian are the tonics suited for building up debilitated animals.

Choking.

When sheep are choked the attendant should not be too hasty in passing a probang if a suitable one is to hand, but try two or three times with a little linseed oil. A piece of new manila rope that has not been wet makes a suitable instrument for passing down the gullet if the obstruction cannot otherwise be moved. A green willow stick, with a bag of linseed firmly tied to it and dipped in hot water makes a useful temporary tool for this job.

Loss of Cud.

This is a symptom rather than a disease in itself. No other sign of disease may be discovered in some cases, and it is assumed then that acidity of the stomach is the cause. An aperient dose with table salt and sulphate of magnesia, equal parts, appears to have the desired effect. Sheep do not tolerate such large doses of salt in proportion to body weight as do cattle, and one-third of salt will be a safer quantity in the less robust.

Hoven or Tympany.

Sheep get blasted by the same means as cattle, and relief must be quickly given. The flank may be punctured midway between the hip and the last rib on the animal's left side. Linseed or olive oil with ten to forty drops of carbolic acid should be administered by the mouth. Careful feeding for a time will probably restore the distended rumen to its normal powers.

PLENALVIA OR DISTENSION OF PAUNCH.

Instead of gas, the rumen is sometimes distended with food, which has accumulated and more or less paralysed the stomach.

Symptoms.—Enlarged flank, more particularly on the left side. Hard and compressible, giving a dough-like feeling to the hand, and slowly refilling as compared with tympany or hoven, which see.

Treatment.—Frequent small doses of salts, ginger, and nux vomica, in warm ale or cider, alternated with linseed oil. Aperients act much better when accompanied with the cordials or true stimulants named. Mustard, pimento, caraways oroth er cordials may be employed by way of change. A veterinary surgeon will operate through the flank and take away two-thirds of the contents under the nearest aseptic conditions obtainable; but sheep do not bear this ruminotomy so well as cattle, and it is a rather desperate resource.

FARDEL BOUND.

The impaction and stoppage commonly attributed to the fardel or maniplies is often in another compartment, and the leaves of the fardel are normally dry as compared with the contents of the other sections of the stomach. Symptoms vary from failure to eat and a grunt or grinding of the teeth, to staggering and mental derangement. Purging, bleeding, and change of food is the treatment needed.

WOOL BALLS AND CONCRETIONS OR STONES.

The felting together of wool and undigested fibre, mixed with particles of earth and sand, occurs from time to time in flocks, and if the history of the flock is carefully inquired into it is found that during a period of short commons the animals have devoured much woody and innutritious material, or from indigestion arising out of other causes have consumed the things of which the balls are composed.

Symptoms are dullness and loss of appetite and standing about alone in a more or less dazed or absent-minded manner, with, later on, heaving of the flank. Bezoars are found in the fourth stomach or abomasum; they are reddish in colour, and with a smooth velvet-like surface.

Wool-balling in Lambs.

This is not quite the same thing as referred to above, but often of serious import to the lamb who acquires a taste for wool or by accident takes it from about the udder of the ewe while sucking. It entangles the curd in the first stomach and leads to flatulent colic, constipation, and often to death.

Treatment.—Ewes should be clatted or close clipped about the region where lambs are liable to take in stray strands of wool. Adults as well as lambs should be given repeated doses of oil, such as castor oil first, and afterwards linseed oil, as these things enable the patient to pass out the bulk of the wool, and they help break up a tangled curd. The concretions of the adult which have become hard and have a nucleus of some foreign body, as a bit of wood or metal, are not likely to be broken up by any treatment, but when of no great size are not infrequently passed in the dung as a result of an aperient dose. If then, members of a flock are known to have these obstructions, it may be well to give a dose in time.

Inflammation of the Stomach or Abomasum.

The fourth compartment of the stomach is often referred to as the true stomach. The abomasum is its anatomical name. Inflammation of this organ was formerly attributed to a variety of causes, but it is now generally recognised to be due to the presence of myriads of parasites of small thread-like form. Strongles of several kinds, outbreaks of which have more than decimated the lamb flocks in autumn and winter in districts as wide apart as Lincolnshire and Somersetshire.

Symptoms.—Scouring of a persistent kind, followed by wasting and loosening of the wool and the usual symptoms of ill-doing. There is what may be called a negative symptom which differentiates this trouble in the fourth stomach from inflammations caused by other things. The lamb or sheep does not seem ill in itself, and gives one the idea that if only the purging could be arrested the animal would be well. The writer does not expect the novice to understand this differentiation, but the man familiar with sheep will appreciate the distinction, and it may assist in diagnosis. Diagnosis is indeed not difficult if one makes a constant practice of examining the dung. Naked eye observation should not be sufficient, as the parasites are often infinitely small and a magnifying glass or pocket lens will demonstrate the presence of thousands, where none could be seen without it.

Treatment.—It is most unsatisfactory we must admit that no really successful treatment has been discovered. There are many drugs which will destroy the worms outside the body, but they fail inside the host, because the presence of the parasites causes a thick layer of tenacious mucus to form, and this provides an almost perfect protection for them. If success is ultimately attained, it will probably be by some alkaline washing of the stomach and thinning down of the mucus before sending in such doses of tobacco as the patient can bear, if sustained by stimulants. Arsenical tablets and turpentine in linseed oil have some little effect in checking the increase of the worms ; and the administration of salt, iron and Peruvian bark helps to sustain the strength of the animals while awaiting the time of their departure. In the matter of preventive treatment, we may take a hint from the discovery of the Grouse Commission in the case of another worm of the same family, the *S. pergracilis*. This parasite passes ova from the infested bird which develop into motile embryos capable of reaching the tips of the heather in four or five days, from which they are taken by the birds. There is therefore a period of migration, when salting and rolling the land must be the means of destroying myriads of strongles of the kind which infest sheep. (See Husk or Verminous Bronchitis.)

Scouring Caused by Worms.

Scouring or diarrhœa has been several times referred to in other sections of this work, and among the causes strongylosis or infestation by minute thread-like worms mentioned. Another and most serious cause of scour is

Tapeworm.

There are many tapeworms recognised by the helminthologist, but the commonest offender is the Tænia expansa, and it often attains to a great length, while vast numbers afflict a single animal. Lambs are preferred as hosts, but adults do not escape. All tapeworms pass through three stages, the egg, the hydatid or cyst, and the mature worm. A variety of intermediate hosts are known. Soft molluscs are among them. Where many rabbits are found, tapeworm is gene ally present among the flocks; where dogs overrun pastures, gid is prevalent (see Gid, sturdy or turnsick). For the full history of the tapeworm the reader will consult such works as Pneumann's " Parasites and Parasitic Diseases." We are more concerned here with the methods of prevention and of treatment at present practicable.

Symptoms.—A previously healthy flock begins to scour and to do badly. Other causes do not appear. The dung should, as advised elsewhere, always be examined, and not only when disease appears, but as a matter of habit. Small segments of tapeworm or proglottides as they are called, will surely be found if carefully looked for. On some days it may be necessary to make quite a search, but on others, white specks will be so numerous as to attract the man with any powers whatever of observation.

Treatment.—All animals intended to be physicked for worms should first be fasted. The period will depend somewhat upon the kind of food already taken. From midday until the next morning is not too long to shut up the verminous flock before giving the dose. Tablets consisting of arsenic, iron and sulphur, and tobacco, are found fairly effective in getting rid of tapeworm, especially in lambs, when the parasites have not been long in possession, and their suckers have not been embedded in the membrane and held fast by adhesions (the reason why human and dog tapeworms of long standing cannot be ejected). Turpentine in linseed oil is another remedy, used with more or less success. Experiments made by the Transvaal Government veterinary officers go to prove that much larger doses of arsenic are tolerated by sheep than is generally supposed. Similar experiments are much needed in this country as a more effectual vermifuge would be found in the greater dose administered at one time. From one grain for a lamb to two grains for a sheep is about the usual quantity deemed safe in England, and lambs can be treated at eight weeks old and upwards.

The Rot, Fluke, Fascioliasis.

The disease known by the above terms, and some others in different districts, is caused by a sole-shaped worm which finds its ultimate home or destination in the channels or so-called ducts of the liver. It passes through several stages, during one of which it lives in the body of a water snail, but the mature parasite, as above stated, is destined for the sheep's liver, where it causes trouble in proportion to its numbers.

Symptoms.—It is generally said that during the first few weeks of the invasion of the liver, the host does well. The stimulation of the liver and increased flow of bile aids digestion and the animal thrives. Then follow febrile symptoms, a general loss of condition and ill-doing, with a disposition to chogging or dropsy under the jaw, wasting of the loin muscles, and dropsy of the belly, wool coming out, short-windedness, general weakness, and pallor of the membranes. When the razor-backed stage is reached it is very little use to keep the animal over to another season, but if so kept, there is improvement. It does not, however, pay to keep rotten sheep.

Treatment.—Early recognition of the disease will save great loss. If on the first suspicion the worst animal is destroyed, and its liver found to be extensively invaded by flukes, it may be assumed that others are similarly infested, and the flockmaster will seize the chance of butchering during that early period when we have said the sheep thrive. This hardly applies to the ewe flock, and it is often a difficult matter to decide what is best to be done. If flukey sheep are to be kept, they may be given salt and iron and gentian and be caked in the hope that good nourishment and tonics will keep them from going back in condition until the ewes have yeaned and the next year's plans can be laid. The position of the parasites in the liver places them beyond all reach of destruction by drugs or any direct destroyers. With good feeding and constitutional vigour, a moderate number of flukes will be tolerated. Ditches, where snails haunt, should be thrown, and lime covered over the heaps, and pastures salted and dressed with soot. (See Inflammation of the Abomasum, also Husk or Verminous Bronchitis.) Such dressing of the land has been found to pay well, because many other kinds of parasites are also destroyed.

CHAPTER XXXII.

DISEASES OF THE URINARY ORGANS.

INFLAMMATION OF THE KIDNEYS.

Inflammation of the kidneys occurs sometimes as an epizootic in lambs, and is probably caused by an organism as yet unidentified. Absorption of mercurial and other preparations may be held to account for a few cases, and in the newly shorn chills may easily be the cause.

Symptoms.—Arching of the back and grinding the teeth, and frequent attempts to urinate without passing any water, or only a very little, and that extremely high coloured.

Treatment.—Mustard to the loins, a laxative dose of linseed oil, and twice daily, ten-grain doses of salicylate of sodium in water. Red water is not a disease of the urinary organs; they merely act as channels. See Blood Diseases.

KIDNEY STONES.

These are sometimes the cause of inflammation; more often they are not discovered during life. Nothing can be done by way of treatment; or nothing that would be worth while from the economic point of view.

STOPPAGE OF THE URINE.

This trouble is practically confined to the male. A block may occur at any portion of the tortuous urethral canal, but in practice we seldom meet with calculi of any size or high up, but frequently have trouble with rams fed too largely on "roots," and more particularly turnips; the deposit being of a sandy or sabulous nature near the end of the penis, which is corkscrew-like and, from its resemblance to a worm, called the vermiform appendix.

Symptoms.—Inability to urinate, or dribbling of urine, and interrupted flow, with signs of discomfort, and desire to pass more.

Treatment.—The ram should be turned, and the penis and sheath carefully examined. In many cases the finger-nail will enable one to scrape away the sandy matter, and little crystals or mixtures of salts and mucus. If, however, the vermiform appendix is and has been blocked for some time, and for some distance, mechanical removal of the offending material becomes very difficult, and a cure improbable, and operation is desirable. The appendix may be cut off, and bleeding arrested by tincture of iron, or other styptics, and healing looked for in a few days. It may be well here to say that the notion of impotence or sterility so generally held by shepherds has been frequently disproved, and that rams have been as fertile after operation as before.

Leaking of Urine from the Navel.

Lambs, like other young creatures, are liable to leaking away of the urine through the urachus or fœtal passage. They should be examined in order to ascertain if the proper outlet is available, and if it is so, the umbilicus may be sealed by tying the short navel string, or picking it up with a needle if too short to get hold of with the fingers. A touch of bluestone may be used. Any blocking of the outlet will generally lead to the passing of urine by the proper channel.

Diseases of the Generative Organs.

These would be few if we could exclude those incidental to parturition. The ewe flock and the lambing-pen are the chief anxieties of the sheep-farmer. Some of these are due to our interference ; as when bad effects follow castration. This necessary operation may be considered here, just to point to the methods of avoidance of them.

Castration.

There are several ways of emasculating the sheep, and each succeeds in the particular district in which it is practised. The Down shepherds prefer operating within the first three weeks, while the organs are very small. The method is to pull down the purse, and cut off about three quarters of an inch with a sheer steel rough-edged knife, and draw each testicle by means of the teeth until the cord breaks. Inexpensive instruments are made for the purpose, and are preferred by most people, but the old plan is the best. In other districts lambs are allowed to grow to two or three months old, but their tails have been docked while quite young. Docking, as well as castrating, is done at the same time in many flocks, but an interval should be allowed, as hæmorrhage from the tail is generally greater than from the broken and contracted arteries of the testicles. Late lambs are not generally drawn, but the cords divided by the hot iron. A greater mortality follows late cutting, and the monetary loss is, of course, larger. With the very young it is not usual to take any antiseptic precautions. With older lambs the green castrator's ointment is used in the teeth of the clams as well as upon the wounds of the scrotum before releasing the animal. When rams are castrated the risk is considerable. There is a disposition to clotting and blood-poisoning from the presence of the decomposing mass. To avoid this, many operators divide the purse with the hot iron, and do not employ a knife at all, and freely dress with the verdigris ointment alluded to as castrator's ointment. It is made as follows : Verdigris, one ounce ; Venice turpentine, three ounces ; lard, ten ounces. The lard and turpentine are melted together at a low heat, and the verdigris stirred in while cooling.

Troubles with the penis will be found under diseases of the urinary organs.

ABORTION.

This may occur as the result of hounds running through the ewe flock, or dogging, or any other source of fright, and quickly follows the event. This sporadic or accidental abortion or premature lambing may affect a large number of ewes, and at the same time within a few days.

INFECTIOUS ABORTION.

The infectious variety takes toll here and there at different times and at various periods of pregnancy, and there is no history of fright or accident.

Symptoms are much the same in both, and individual treatment of little avail ; the great thing to be done is to arrest its spread by early separation of aborters and all suspects, and, if housed, by disinfection and removal. It must not be forgotten that the tup may be the means of spreading the trouble.

PARTURITION TROUBLES.

Difficult lambing is due to a variety of causes. Some of them beyond our control, and generally recognised too late, as when water-bellied flocks have been fed on too many turnips or other roots, and the ewes are debilitated from this or other causes such as parasitism. The breeding flock should be kept in fair condition —not fat and not poor, and, above all, not debilitated.

EVERSION OF THE VAGINA.

This happens often before lambing, and is mistaken for coming out of the womb—a thing impossible without the ejection of the lamb. It may be caused by a water-bellied condition and pressure within which the weakened parts cannot sustain, or it may be a symptom of the abortion bacillus at work.

It is often possible both to repose it and to keep it in place by

Q

the application of a West's clamp or by large stitches, and ewes have lambed successfully, despite so serious an accident before yeaning-time, but one to which this has happened should not again be run with the tup. The subject will be again referred to as a *post partum* accident.

MAL-PRESENTATIONS.

The normal presentation for a single lamb is that of head and fore legs, the feet being a little in advance of the muzzle. In the case of twins, it most often happens that the second to be born comes with the hind feet first. Neither of these positions should present any difficulty unless the lamb is abnormally large or the ewe unnaturally small, or the labour a dry one, as it is called. A dry labour is the result of previous leaking away of the amniotic fluid or contents of the lamb-bag. It may have dribbled away a little at a time, or the bag which should come first and fill the passage and prepare the way may have prematurely ruptured without the succeeding labour pains being energetic enough to expel the fœtus. Then the normal lubrication is wanting, and greater striving becomes necessary in order that the lamb may be born. The observant shepherd will have noted a leaker, and perhaps a breaker, and will then give that early assistance which is, in the majority of cases, to be deprecated, for it is always a temptation to the accoucheur to act too soon rather than to assure himself of a correct position of the fœtus and wait for the normal efforts to eject the lamb in due time. The man who has attended foaling mares must reverse his views. The mare must give birth quickly or the foal will be dead. The ewe may be quite a long time in labour and bring forth alive.

Manual Aid.—The ewe that is seen to strain without result should be secured with gentle, but firm, handling, and the position made out by examination with hands prepared by dressing with carbolic oil or a watery solution of some other disinfectant. The accoucheur should have well pared nails and washed hands that will not be likely to abrade the delicate membrane of the vagina. Some comparatively trifling obstacle may be hindering parturition, such as a folded back foot, or depressed chin, or head slightly sideways, and be capable of easy alignment or correct presentation. If such is the case, the busy shepherd will, for the time, do no more, but will keep the ewe under observation. Much more serious difficulties are met with, such as a turned back head, or one leg forward and one back, or both limbs behind. In breach presentations the tail may be felt, but no limbs; the feet being below the brim of the pelvis, or the loin may be against the passage, and neither head, limbs, nor tail can be felt, so that the would-be deliverer is puzzled where to begin. We have known two, and even three, feet presented, and they have not all belonged to the same lamb. In the space at our disposal it would not be possible to describe all these mal-presentations, and give directions as to the means of overcoming the difficulties, but we can concentrate the experience of many cold and miserable nights in the lambing-pen into a few earnest words of advice to the beginner to remember that *there is plenty of room where the lamb came from.* Almost all the difficulty that the beginner meets with is through want of remembering this fact. He tries to get his hand into a narrow passage already full, and finds he cannot work. He fears to let a foot or other part out of sight. If he will provide himself with a few cords with one, two, and three knots in them, by which he may presently know to what they belong, he can secure a part without fearing never to see it again. He can then push back the unborn one, and commence afresh to get the parts into line. If he can see nothing, and can only feel a living wall (as when the back is presented), he can push back with better hope that some other part will come with the next labour pain. One other word to the accoucheur, and that is, pull only when the ewe herself is striving, unless she is spent and has given it up. Then he must hasten, and then only, but he should give a rousing stimulant before employing force.

COMING OUT OF THE WOMB.

Following on difficult or prolonged labours the womb may become everted when the after-pains occur. This is a serious accident, and only prompt measures will succeed in saving the animal's life. The parts should be carefully cleansed and dressed with a strong carbolic mixture, such as one part of acid to fourteen of olive oil, then carried in with gentle, continuous force, and secured by a West's clamp. The ewe should be given two to four drams of laudanum with half an ounce of sweet spirit of nitre and made comfortable on a dry place.

COMING OUT OF VAGINA AND OF BLADDER.

In some cases the womb will hold, but the vagina in part is everted after the strain of parturition, or the bladder is turned inside out. A veterinary surgeon will be able to replace the bladder, and we have known some shepherds too who have succeeded, but it needs very nice handling, and cannot well be described. When replaced, the same sedative treatment as for coming out of the womb is recommended.

NOT CLEANSING.

Retention of the placental membranes is usual in abortions and very general in premature births. It is not rare when the full time has gone. The membranes sometimes get caught in the closing of the neck of the bottle, as the *os uteri* is sometimes called ; or the roses (*cotilydons*) by which the membrane is attached to the womb have not matured and let go their hold.

Treatment.—Interference on the first day is dangerous. On the next, gentle traction may be used, or, on the third, considerable pull may be allowed, as the danger of hæmorrhage has passed and a new one begun in the form of septic poisoning if the decaying mass is much longer retained. Winding between rough sticks is better than using the hand. A small bag filled with sand may be attached. Syringing out with a weak disinfectant such as a scruple of permanganate of potash in a pint of warm water is a good thing, and care should be taken to lift the membranes so that the part upon which they have been lying gets properly dressed.

INFLAMMATION OF THE WOMB.

This dangerous complaint follows on difficult labours or any abrasion of the genital membrane through which septic matter has gained access.

Symptoms.—The ewe goes bad, and humps her back, and neglects her lamb (if she has one living), ceases to feed, and has a high temperature and quick, thready pulse. When the lips of the vagina are parted, the membrane will be seen to have a dark-red or purple hue, and there will be more or less sanious discharge.

Treatment.—A dose of linseed oil, gruel with stimulants to hold her up in the fight against prostration. Syringing out with a mixture of four per cent. carbolic acid, seven per cent. glycerine, and ninety per cent. of warm water. Doses of five to ten grains of quinine twice a day have given good results.

GARGET.

This trouble may be due to the udder coming in contact with frosted ground or from blows from the lamb's head or other outward violence. East winds setting in suddenly cause it.

Symptoms.—Uneasiness and walking wide behind or standing with hind legs apart. Refusing the lamb. Swelling of the bag, loss of appetite, high temperature, quick breathing.

Treatment.—A strong aperient, such as a dram of aloes or more, half an ounce of nitre, and three ounces of salts, dissolved in a gill of warm water. Temperature runs higher in this complaint than in any other, and larger doses of aperient medicines are both tolerated and needful. Fomentation of the udder with warm water and gentle rubbing in of camphorated oil is the treatment. No strong " oils " or liniments should be used.

INFECTIOUS GARGET.

A terrible scourge is infectious garget. It takes the ewe very suddenly, and is often not correctly diagnosed until too late to save the quarter from death.

Symptoms.—When seen early, are the same as above described, but the tendency to death of the structures involved is so rapid that it is usual to find the udder already purple to black.

Treatment.—Not the individual, but the flock must be our first consideration. To get her away from the others before the discharges have a chance to distil venom. If caught in the early stage the rather desperate measure may be taken of injecting with a hypodermic syringe, at several points, a solution of perchloride of mercury of the strength of one in fifteen hundred of water. We have known one in one thousand to be used. Ewes are so very likely to die that ordinary caution is set aside in attempting a cure. We are disposed to credit very large doses of quinine, such as twenty to thirty grains with good effect. The bag should be liberally dressed with the camphorated oil previously recommended. If the quarter sloughs out and the ewe is saved, she should be made up for the butcher.

CHAPTER XXXIII.

DISEASES OF THE NERVOUS SYSTEM.

The nervous system of the sheep is not as highly developed as those animals in more close association with man. The dog first, and the horse next. Timidity is not a sign of a developed nervous system Most of the brain and spinal cord diseases are parasitic in origin, and those that are not traceable to such causes are still obscure.

GID, STURDY, TURNSICK.

This commonest of brain troubles is due to the presence of the bladder caused by the embryo tapeworm which has been taken into the stomach, pierced the wall, and gained access to the blood, using its instinct and its hooks to anchor in the brain and set up the formation of a bladder upon the inner aspect of which the heads or first portions of future tapeworms form.

Symptoms.—Giddiness, carrying the head to one side and falling over, or spinning round and falling over. After a short time the giddy sheep gets up again, looking a bit dazed, but soon appears to be in its usual health.

Treatment.—Puncturing the bladder if it can be located. It is always more or less risky, but is often done with success. A soft, place can sometimes be felt in the skull, and into this a fine awl driven, or a hypodermic needle of stout proportions, which will also draw off a large part of the fluid. Another plan, that adopted by Hogg, and called by him " wiring," is to push up the nostrils a long wire or knitting-needle into the bladder, which he had a great knack of finding. "If," says Hogg, " the brain does not inflame, the sheep will be better." Piercing from this aspect (the nasal chamber) allows of drainage, and is preferable on that account. Any puncture of the bladder has a good effect, but the reader will see how much greater the result of aspirating, by the springe as above stated. As to whether one will undertake this operation or not will depend upon whether a breeding ewe is affected or a nearly fat hogg or sheep destined for the butcher. Aperients and bleeding, any practice which lowers brain pressure acts favourably, and if the period of maximum enlargement can be passed, the sheep will recover, as the contents of the cyst begin to contract, and the brain has reached a certain amount of accommodation; some of this by actual absorption or wasting of the brain substance. The means of prevention are referred to in the space devoted to tapeworm, which see.

FITS.

Epileptic or epileptiform fits are sometimes seen in sheep, more particularly in lambs brought from bare hillsides on to rich pastures.

Treatment.—Bleed, if the symptoms are very urgent. If not, the remedy will be found in purgatives and reduced rations, or keeping off rich pasture, and hay feeding first, so that the animals do not eat so much. They get increased ability to profit by rich food after a while, and then the fits cease.

LOUPING ILL.

Very little is known about the cause of this disease, but change of pasture or removal from danger areas at particular seasons of the year seem to be the most important steps to take.

TETANUS OR LOCKJAW.

This follows on injuries, and is caused by a specific bacillus. Lambs, after castration and docking, are the most frequent subjects. Land known for it should be avoided at lambing-time. Fresh-cut or docked lambs should be put on clean, dry straw until the wounds are sealed and dry.

Treatment consists in removing the clots of blood from the wounds or picking off the dry scabs and disinfecting to the bottom, as by this means the supply of toxin is cut off, and the animal recovers if the " dose " has not been too large.

CHAPTER XXXIV.

DISEASES OF THE EYE AND SKIN.

Sheep are not specially prone to diseases of the eye if we except those eye blights which from time to time come over lamb flocks and sometimes older sheep. Driving east winds and the effects of summer dust when flocks make long journeys by road result in common inflammation.

CONJUNCTIVITIS.

Inflammation of the membrane covering the eye causes pain and lost time in grazing.

Symptoms.—Partially closed eyes, standing about and not feeding, running of watery tears down the face, and similar overflow down the lachyrymal duct, and from the nostrils. At a later stage the discharge thickens, and around the corners of the eye matter dries in gummy form. If neglected, a white film is liable to form, but it usually clears up after a time, and beyond the annoyance of the early symptoms above described, the sheep does not suffer, and can find and select its food if temporarily blind.

Treatment consists in placing the flock under more favourable conditions as to wind and protection from the causes. Better food and tonics are sometimes desirable, as a catarrh of a bad type prevails in some seasons and reduces vitality and so checks the development of the young and prejudices the ewe flock.

EYE BLIGHT, INFECTIOUS.

This appears to be due to a specific organism and spreads rapidly among lambs of the previous spring. It comes in the late autumn months, November being the most frequent month for its appearance. Mild antiseptics such as two or three grains of sulphate of zinc to each ounce of distilled water, dropped into the eyes daily, have proved curative, and no doubt assist in clearing up the clouded membrane. If this fails, a solution of nitrate of silver of similar strength may be used. The first to be affected should be removed, and a daily watch kept with a view to separation of any others that may be weeping, as it is very infectious. The inflammation caused by insects may be treated with the same lotions as those advised above, and shade should be given as the most important part of the treatment in all inflammations of the eyes.

DISEASES OF THE SKIN.

These are of two kinds : those coming from within, as eruptions ; and those induced by the presence of parasites or external injury. The specific eruptions caused by sheep-pox we need not consider, as they call for immediate notification. The sheep is ill before the typical eruptions take place. They then pass through certain recognised periods of papulation, vesication, pustulation, and desquamation, by which they are distinguished from simple eruptions.

ECZEMA.

Upon the uncovered portions of the skin, as under the thighs and arms, eruptions appear as the result of indigestion, or foods unsuitable in character.

Symptoms.—The animal appears dull, and moves with some discomfort or soreness, which attracts attention. There may be some elevation of temperature and indifference to food in the formative period. The redness passes on to elevation of the surface or pimply spots which presently contain a watery or milky fluid. The movements of the animal are liable to break these pimples or vesicles, and then a confluent sore is discovered.

Treatment.—A dose or two of salts, with a daily dram of bicarbonate of potash or soda, and the application of some lard (without salt) or lanoline ointment—any simple emollient will usually suffice. The chief matter in treatment is the discovery of the cause and its removal.

DROPSICAL SKIN.

Fluid swellings under the skin, as in rot, and other constitutional conditions, call for treatment of the disease itself. They are but local manifestations of a systemic condition, and not properly diseases of the skin.

SCAB OR SHAB.

This is caused by a parasite of the mange class, and might be described as sheep mange.

Symptoms.—Rubbing and coming out of the wool.

Treatment.—There is no great difficulty in destroying the parasites that give rise to scab, but there is often much trouble in preventing its spread, and tracing the lines of communication so to speak.   A flock may be properly dipped, and yet fresh cases occur. Every possible rubbing post should be dressed, and any sheep showing irritation caught up and a little scurf taken from the outside border of the rub and examined under a magnifier, for broken bodies of the mites.   There are many good dips on the market, and it generally pays better to buy them than to make preparations, unless one farms in a tobacco-growing country.

LICE, KEDS, TICKS, ETC.

All the parasites which bother sheep can be equally well disposed of by dipping, but there are occasions when only limited areas are affected, and the weather may be unsuitable for dipping, and then we may resort to mercurial or " sheep " ointment for rubs over the parts infested.

LEG OR FOOT MANGE.

This is a much more difficult disease to cure than that affecting the woolly parts, and so well known as scab.   It is due to a bacillus, the Necrosis bacillus, and may be communicated from the rubbing of the face on the leg, or by the leg to the lips.   It is considered under the heading of pustular dermatitis or lip and leg mange, which see.

FLY OR MAGGOT.

A cause of frequent annoyance, and in some seasons, serious injury, is " the fly."   The kind which most afflict sheep are most found under trees, as they seek shelter from rain and are seldom found at any distance from such retreats.   They are attracted to milking ewes and to such sheep as have scour and the hind parts soiled.   Any sores such as rams get about the head as the result of fighting, or cuts incurred in shearing, offer invitations to the *Sarcophagus carnaria.*   The injuries done to the feet by foot-rot or that necrobacillosis referred to in other parts of this section are made worse by the fly, which deposits its eggs, and these quickly form squirming masses of maggots, which live upon the yolk, and cause great wounds as well as small.

Symptoms.—The commonest sign of fly is shaking of the tail and rapid movements, as if stung by pain.   Any irritation displayed in this manner or by rubbing (see Scab) should cause the shepherd to examine the animal without delay.

Treatment.—Many sheepmen carry a lump of fly stone in the season when these pests are about, and apply it with excellent effect so far as killing the maggots instantly are concerned, but this perchloride of mercury or fly stone, or corrosive sublimate, as it is also called, destroys the wool roots, and leaves a blemish where no wool will again grow, and more or less of a scar and detriment to the pelt as well. A better plan is to use a rather strong mixture of carbolic acid and seed oil, such as one part in twenty. This kills the maggots, and spreads in a manner to cover stragglers, and promotes healing. While it lasts, it is a deterrent to other flies settling on the wound.

Prevention.—In fly areas it is better to anticipate those sultry days when ovipositing may be expected. A fly powder properly applied will give protection for some time, and may be repeated. The early morning, when the fleece is dewy, is the best time to apply it. At other times, a watering-pot with a fine rose may be used. The fleece should be parted along the back to receive the powder, and special attention given to the parts round the root of the tail. The following is a suitable mixture : White lead, 2 lb. ; red lead, 1 lb. ; sulphur, 1½ lb. ; spirit of tar, 4 oz. The two leads should be rubbed down with the tar spirit ; the sulphur separately treated, by passing through a fine sieve, and then mixed with the leads. The mixture is a little too damp for immediate use through the holes of a flour dredger, and will work better for being first spread on sheets of paper for a few hours. It can then be put away, and will keep indefinitely.

SORE HEADS.

If by neglect the sores resulting from ram fights have spread, the following powder will be suitable for dusting on : Powdered alum, 1 oz. ; Armenian bole, 1 oz. ; white lead, 1 oz. ; chlorinated lime, 1 dram ; prepared chalk, 3 oz. Some bad cases need a cap, of the nightcap pattern, to protect the head until the wound has dried and become unattractive to the fly.

CHAPTER XXXV.

SPECIFIC DISEASES.

Joint Ill, Navel Ill.

Reference has been made to some of these, such as sheep-pox.

These are infectious, either through the open navel or by other portals. In nearly all cases in which the navel is early treated there is an absence of joint ill and of navel ill, but the fact that a few animals will be attacked, although proper precautions have been taken, leads us to suppose that the germs may sometimes enter by the mouth or by abrasions elsewhere.

Symptoms.—Dullness, loss of appetite, or indifference to the teat, increased temperature, accelerated breathing, and a disposition to lie down and remain on the ground. These symptoms commonly precede joint swellings. The latter may not develop at all, but the lamb may die from the pyæmic state of the umbilical cord, which is invaded by organisms which may be said to poison the animal. The navel is moist and swollen, or discharging matter, or the latter may give rise to a swelling because it cannot find vent. The lamb may die in two or three days or linger for weeks. The effects on the joints are secondary, although in a few cases they may be primarily affected. They swell and are very painful, and the young animal cannot stand or kneel to suck.

Treatment.—When once fully developed, treatment is seldom really successful, although better results have been obtained of late years by veterinary surgeons adopting the serum vaccine treatment, which is not yet available to the flockmaster. Also the injection, around the swollen joints, of a four per cent. formalin solution, by means of the hypodermic syringe has been found beneficial. Careful syringing into the navel of disinfectants of three to five per cent. strength in the early period of the attack is thought helpful, and the giving of strong tonic medicines, as a grain of quinine twice a day to a young lamb. Painting the navel as well as the joints with tincture of iodine is also practised. No strong oils or liniments should be employed.

Prevention.—On tainted ground where lambs have previously

suffered, the precaution should be taken of tying the navel-strings as soon as possible after birth. To the tying should be added a rapid painting with a mixture of one part carbolic acid to eleven parts of collodion. This is destructive of the germs, and a fence against their entrance, until the cord has time to dry off, after which the danger is very slight. The lambing-pen should be upon a fresh site each year if possible, and the land salted and dressed with soot, as these substances are inimical to the germs which are known to cause the disease. The practice of lambing in small bartons and in dirty enclosures is responsible for many diseases of ewes as well as lambs. If compelled to use such places, the floor or ground surface should be liberally sprinkled with lime, and clean straw laid down.

Infectious Scour.

This has been referred to in the chapter dealing with diseases of digestion, but belongs more properly to the specific infections. The germs are believed to enter chiefly through the open umbilicus, and this constitutes another reason for adopting the practice of ligaturing and dressing the navel-string with such things as mentioned above for the prevention of joint ill.

Treatment.—This is very unsatisfactory, as sheep, and more particularly lambs, do not respond to the ordinary astringents such as chalk mixture and catechu, or those compounds commonly sold by agricultural or veterinary chemists, and composed of opium, chalk, bismuth, catechu, and oak bark infusions, or decoctions of other astringent barks and roots. An aperient of castor oil should be tried, and followed up with large doses of quinine at one part of the day, and with scour mixture at another. Quinine appears to act well in large doses by inhibiting the multiplication of the germs, as in the case of certain malarial fevers.

CHAPTER XXXVI.

LAMENESS. DISEASES OF THE FEET.

Some foot troubles and diseases affecting the lower portions of the limbs have already been referred to in connection with so-called quittor or necrobacillosis. Many diseases attributed to wet seasons and other causes are now known to be due to specific organisms.

FOOT-ROT.

The old controversy as to the infectiousness of foot-rot is practically dead, as it is now generally recognised that there are at least two forms of the disease, one of which may be regarded as accidental, and the other as infectious. The common or accidental form has a variety of causes. It may affect quite young lambs born upon moist pastures and with no hardening influences upon the feet while exposed to the risks of pricks and injuries from thorns and other foreign bodies. Any breach in the hoof is liable to lead to decay of the horn structure, and this is foot-rot. Older sheep get it from two quite opposite causes, namely overgrowth on grass, and wear and tear by travelling. The crust or outer wall of the hoof grows beyond the level of the sole on moist land where no attrition is provided by road travelling from one feeding ground to another, and this is followed by strain on the wall, and more or less separation. Into the fissure thus made enters dirt and water, and irritation is set up, the result being the production of imperfect horn—foot-rot in a word. Sheep that have been taken off soft ground and driven long distances by road wear away their feet so quickly as to expose the sensitive parts, dirt enters, inflammation follows, and decaying degraded horn is produced—in a word, foot-rot.

Symptoms.—Lameness, kneeling to feed and, if the hind feet cause much pain, loss of flesh and condition.

Treatment.—Many good remedies are well known, and all sorts are advertised, but their success depends upon the man who applies them. The same remedies even will not be equally successful when used for foot-rot arising from the two opposite troubles mentioned. The overgrown or separated hoof will need judicious paring with the knife to get a fair level surface to tread on, and will need that the medicament be gently pressed into, or poured into, the crevices. The worn foot may need protection with a tarred rag for a few days before a caustic preparation is used to arrest the decomposing horn. As with horses suffering from thrush, foot-rot of the sporadic or accidental kind is greatly helped by fair pressure upon the soft and diseased portions in weight bearing.

INFECTIOUS FOOT-ROT.

The peculiarity of this form of foot trouble is that its infectivity is not direct, as between one sheep and another, for the diseased horn has been repeatedly placed between the digits of sound sheep, and there maintained by mechanical contrivances, without infecting the animal. This was taken to be proof at one time that foot-rot was not infectious, but practical sheepmen persisted in regarding it as infectious still. The explanation is to be found in the soil. Where infected sheep have been pastured, clean sheep will become infected. It has not yet been shown that the organisms causing foot-rot pass an intermediate life or period of development in the soil, but there is overwhelming proof that sheep do become infected by the land, and not directly by association with the diseased. Some minor differences in the manner of invasion are recognised, or thought to be, by experts, but lameness is the attractive one, and only sheep kept for experimental purposes would be examined before this characteristic symptom became apparent.

Treatment and Prevention.—The Irish Department of Agriculture, in a leaflet issued on the subject, says : " The sheep owner will be wise if he treats every case which may arise as if it were infectious." The individual handling of a large number of sheep is often difficult, or impossible in some circumstances, and we have to do the next best thing. Sheep can be folded on a half-inch layer of slaked lime, for an hour or two daily, or made to pass in single file through a narrow gateway or hurdled space that will only admit one at a time. The ewe flock must not be handled for foot dressing, as abortion is easily provoked by fright, and for such, the narrow way is less desirable than the folded yard. Lime mixed with powdered sulphate of copper or iron of the cheap commercial quality is found useful for a self-dressing of the kind. Solutions are preferred by many, as being less wasteful and perhaps more effective. A good one may be made of 1 lb. of copper sulphate, and 1 lb. of burnt alum, dissolved in twenty gallons of water. Arsenical solutions are also used, but care is necessary in the subsequent placing of the sheep, or water may be impregnated or grass killed or other animals injured.

In order to dress the feet with either of the preparations mentioned, long shallow troughs must be supplied, and the sides and top hurdled, in order to pass the animals through one at a time, and slowly, to prevent leaping. Arsenical sheep dips are often used in this way, but it is not an economic method.

Individual Dressing.—Flocks that can be turned and handled any way will do much better for having the feet pared carefully and the morbid horn scraped away and cracks cleaned out before applying the dressing. For an ointment, we can recommend a compound of 1 oz. of sulphate of copper, 1 oz. of Venice turpentine, and 8 oz. of lard. The two last ingredients should be melted together at a low heat, and the powder stirred in while cooling. Among the liquid remedies, the old-fashioned dressing of one part of butyr of antimony and two parts of train oil holds its own. Another good mixture is that of one part of carbolic acid in twenty parts of glycerine. This is very penetrating and has other recommendations, but is costly. In some cases there will be found fungoid growths that must be cut away before any dressing can possibly cure. In others, the use of the foot pulls open the cracks. For the latter a tarred piece of canvas or fillis bound round and tarred heavily, answers well. The movements must be confined until the healing process has made progress. It may be said that tar and salt in the proportion of five of the former to one of the latter, will cure most cases of foot-rot in the early stage, and that preference should always be given over those of the caustic remedies, reserving them for the "chronics" and those with outgrowths that must be destroyed. The hoof has a disposition to contract when strong remedies are used, and this may cause lameness, apart from decayed horn.

CHAPTER XXXVII.

MEDICINES.

The practical sheep doctor will give the tarpot the first place in his *materia medica*, because it is both a preventive and a curative agent for a variety of troubles both of the skin, the flesh, and the hoofs, and he can trust anybody to apply a dab of tar, which may save fly-striking and infection of wounds, and preserve injured or foot-rotted feet until more particular attention can be given. Stockholm or Archangel tar is always meant, and the purchaser should be explicit about this, or he may be supplied with coal tar.

Simple remedies are throughout this veterinary section recommended, because they are obtainable almost everywhere ; they are usually safe, and it is far better to have an intimate knowledge of the effects of a few tried drugs than to rely on nostrums which may not be available when wanted.

Remedies have for the most part to be given to sheep in their food, and powders are preferred. They should be intimately mixed in order to give an approximate dose to each animal, and no more food should be given at the time than is likely to be cleared up at one meal—and one dose.

If drenching has to be performed, the doctor or " nurse " will keep dogs out of the way, and get his patient into a corner and secure him with the least possible excitement and noise. The drench may be given in a horn, but a stout sauce bottle has the advantage of transparency, and one can see how much has been given. The head of the patient should never be turned, but held straight—many accidents resulting from neglect of this precaution. The gullet of the sheep is relatively large, and a good quantity is easily swallowed if the lip of the bottle is first pressed against the palate so as to let the sheep know that something is coming, and then he involuntarily closes the glottis to prevent medicine " going the wrong way." With cattle as well as sheep, ninety-nine per cent. of accidents occur for want of this simple precaution.

Medicines should be diluted with plenty of water, although this necessitates more time in administration. Salines act better

R

in comparatively small doses if given with a large quantity of water than do large doses dissolved in only just sufficient water. Two ounces of Epsom salts in a full pint of water will act better than four ounces in half the quantity of water. If a spasmodic cough suggests a drop the wrong way, the sheep's head should be instantly lowered, and the cough allowed to subside before going on with the drenching.

Balls or Boluses.

It is not so generally known as it should be that balls can be given with advantage to sheep where, for instance, it is desired to bring the drug into contact with the parasites which infest the stomachs or the small intestine.

Electuaries or drugs made into paste or of the consistence of butter can be placed far back on the tongue when sore throat or other mouth troubles make it dangerous to give drenches or balls.

Pessaries are wax-like bodies in which active medicaments are mixed, and can be placed in the genital passage to allay straining or reduce inflammation, or as simple antiseptics whose action will be long continued.

Drugs required for a flock often aggregate in considerable bulk, and the flockmaster should buy in one lot for his probable requirements, as all druggists can quote better prices for wholesale than for retail quantities.

The following list of drugs should be kept in stock :—

> Tar.
> Salts.
> Glauber salts.
> Sulphur.
> Linseed oil.
> Castor oil.
> Carbolic acid, No. 5.
> Glycerine.
> Carbolic oil (ready mixed) 1-20.
> Laudanum.
> Sweet spirit of nitre.
> Nitre.
> Corrosive sublimate (poison).
> Tincture of iodine.

The foregoing list comprises what might be deemed the essentials for a flockmaster out of reach of professional assistance or far from a drug store.

To this should be added a few instruments, such as a trochar and canula, with which to relieve a blown animal promptly.

A probang to pass down the throat of a choked sheep.

A few needles and sutures for wounds.

Several of West's clamps, for coming out of the womb (see article on the subject).

A castrating knife, or two, as a dropped knife often cannot be found at the time.

One or more shear-steel shoemakers' knives for tailing.

A pair of strong dental forceps such as are used for human molars, and a pair of " straight stumps " for lamb's mouths.

A pair of pointed forceps for drawing thorns from the tender feet of lambs.

A few sheets of glass-paper and a rasp should be among the things provided for foot cases.

String, tarred cord, and glue bandages can be utilised by a handy man for bone setting.

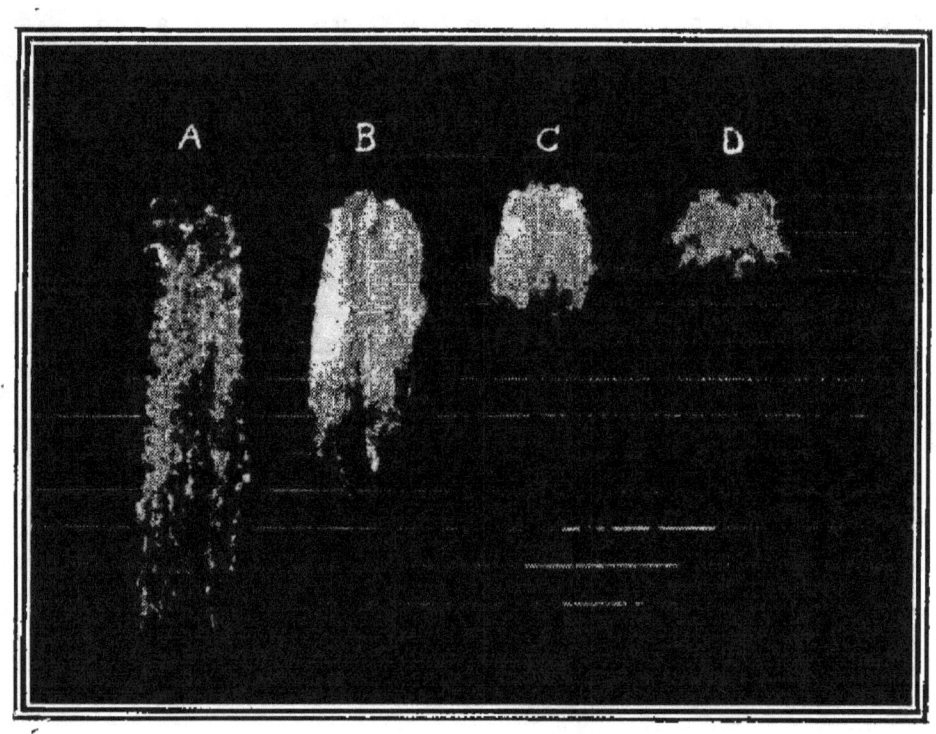

SAMPLES OF WOOL

A, LINCOLN; B, NORTH; C, LONG DOWN; D, SOUTH DOWN.
LINES ON SKETCH ARE ONE INCH APART.

THE OLD NORFOLK HORNED BREED.
The lamb represents a Southdown cross.
(*Professor Low's Illustrations.*)

THE OLD WILTSHIRE BREED.
(Professor Low's Illustrations.)

THE OLD CHEVIOT TYPE.
(*Professor Low's Illustrations.*)

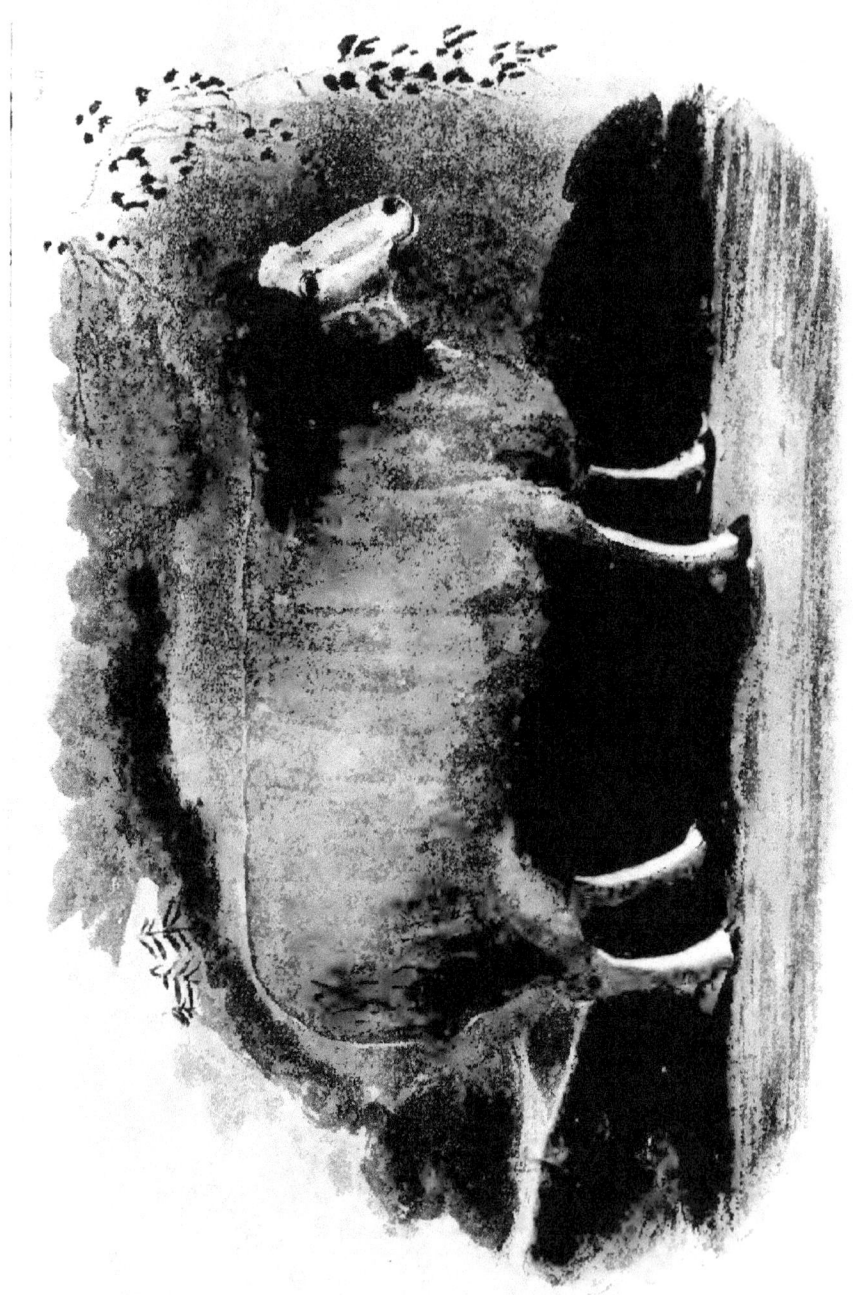

THE LEICESTER.

Described in Professor Low's Illustrations as the New Leicester.

THE OLD MERINO AS DEPICTED IN PROFESSOR LOW'S ILLUSTRATIONS OF THE OLDER TYPES.

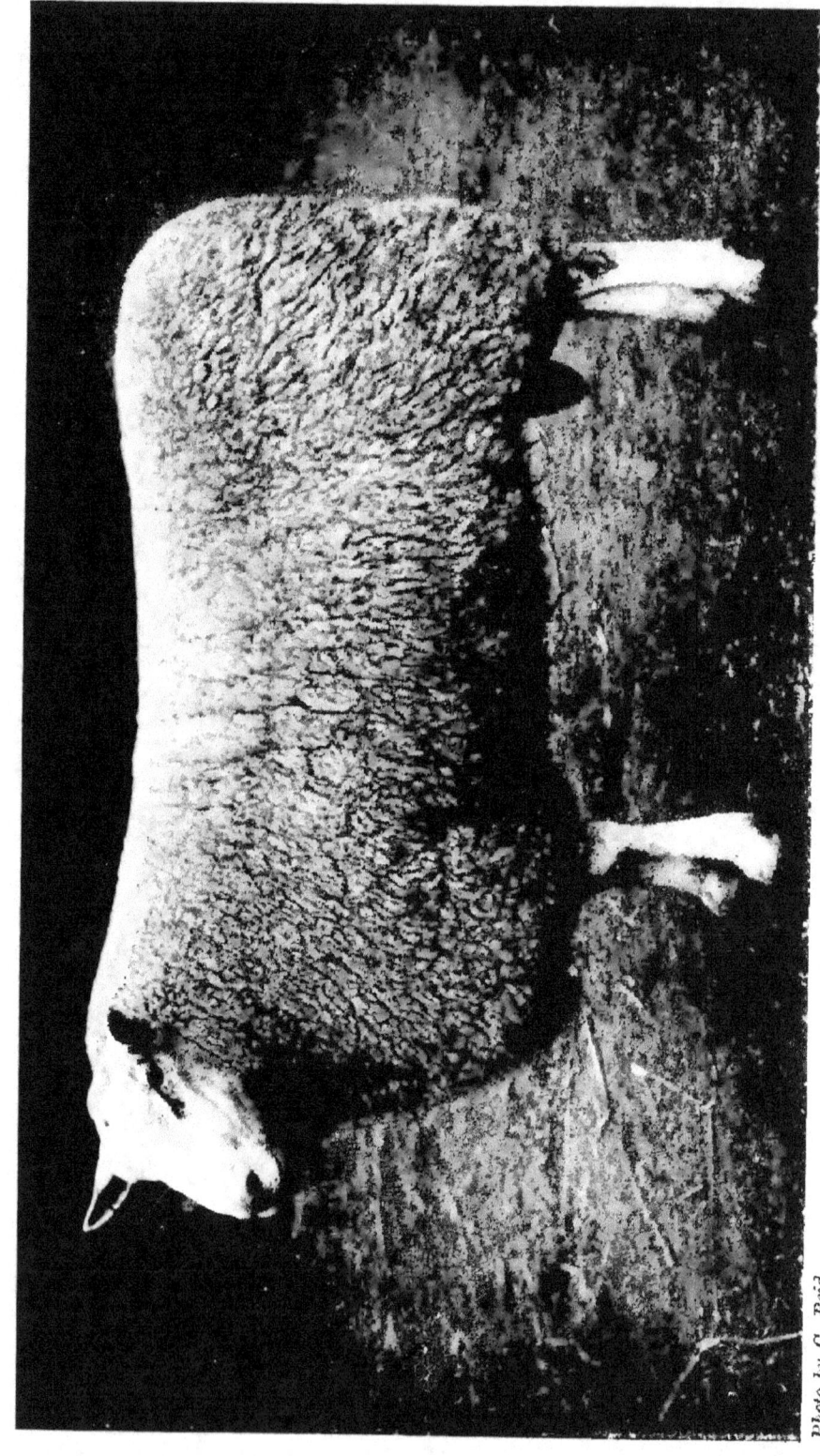

A HALF-BRED BORDER LEICESTER-CHEVIOT RAM.

EXMOOR HORN RAM.

ROSCOMMON RAM.

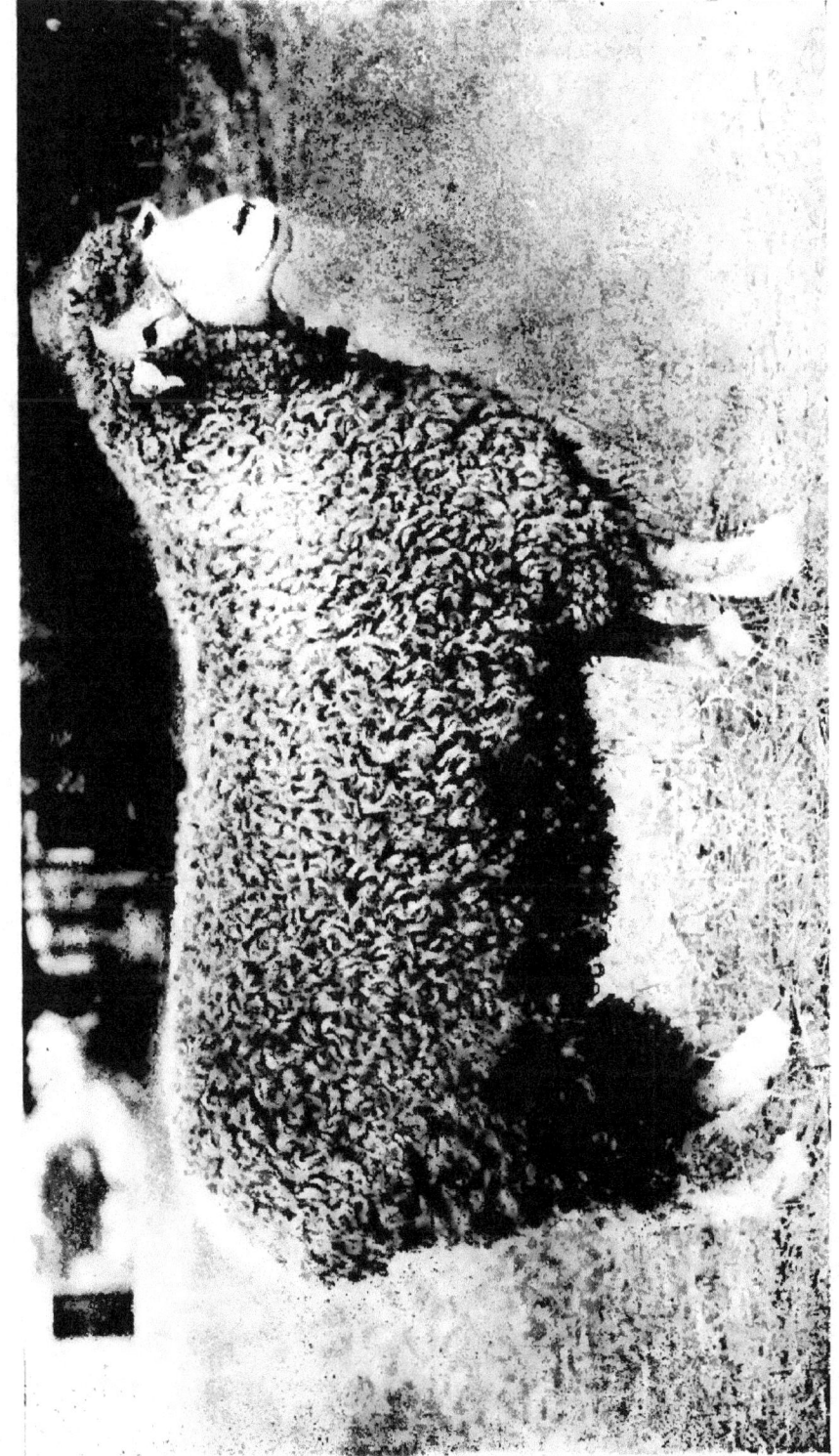

Photo by Sport and General.

SOUTH DEVON SHEARLING RAM.

DEVON LONGWOOL SHEARLING RAM.

Photo by Sport and General.

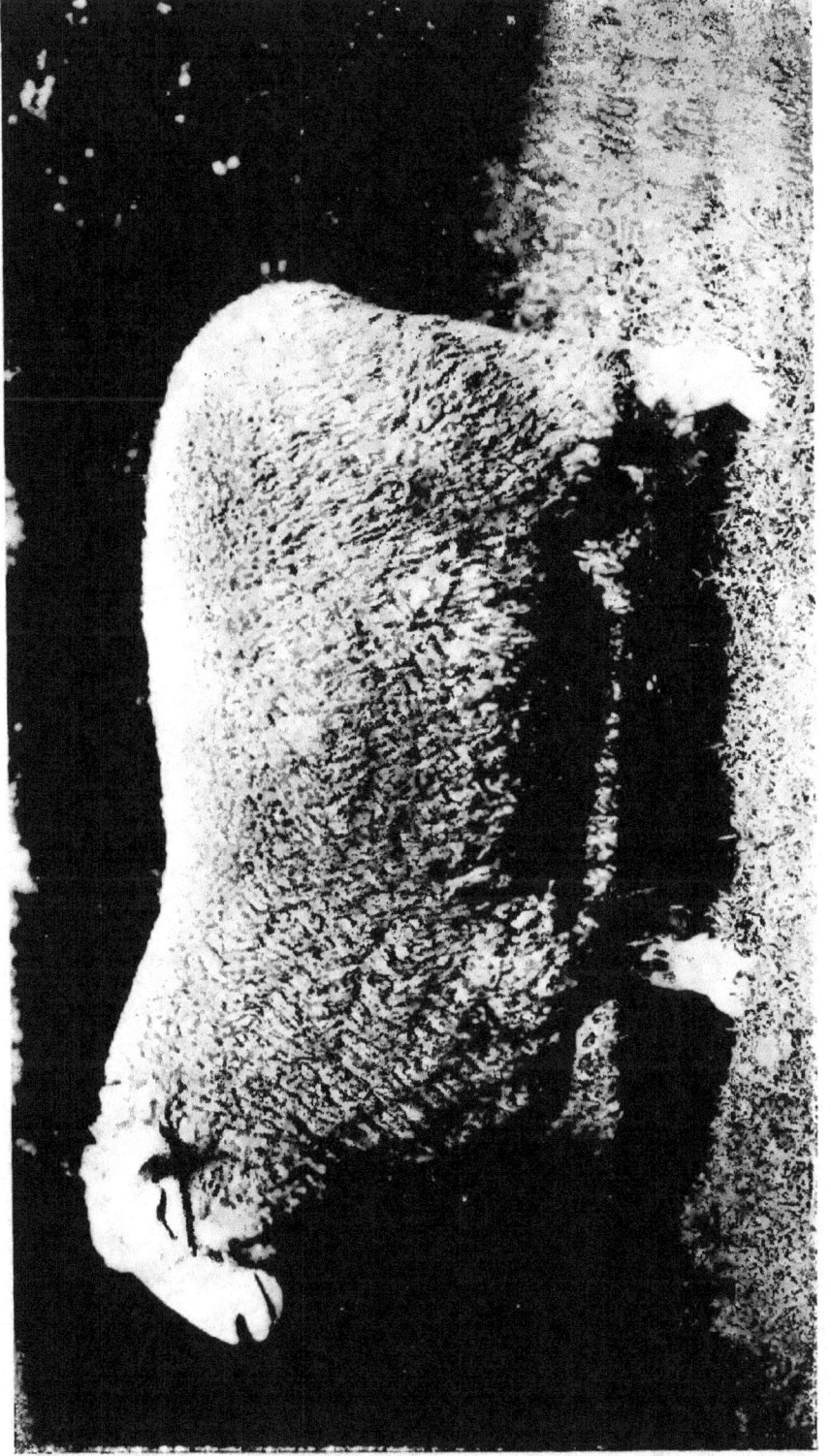

KENT OR ROMNEY MARSH SHEARLING RAM.

Photo by Sport and General.

Photo by Sport and General.

COTSWOLD SHEARLING RAM.

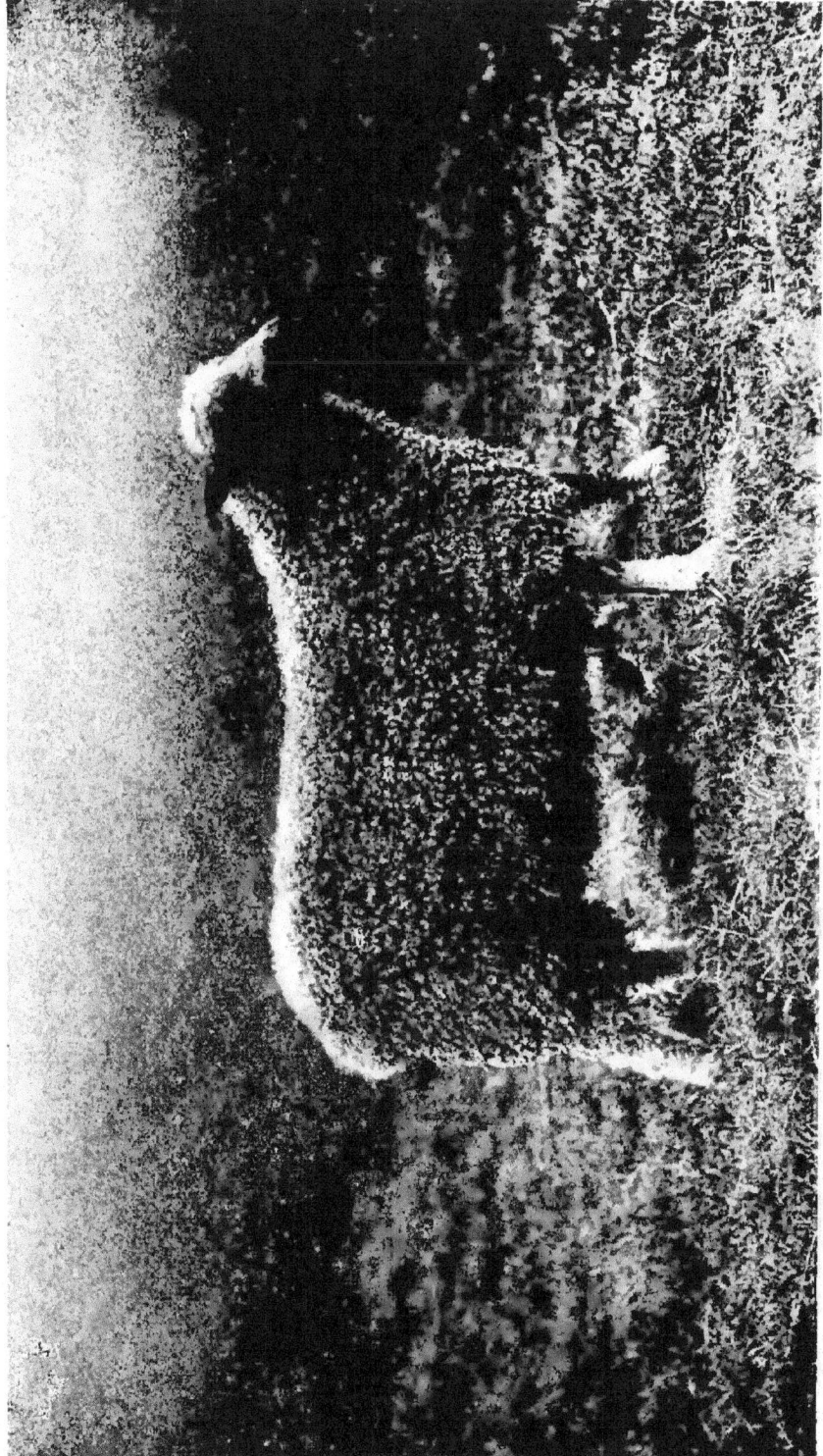

WENSLEYDALE SHEARLING RAM.

Photo by Sport and General.

LINCOLN LONGWOOL SHEARLING RAM.]

+ +

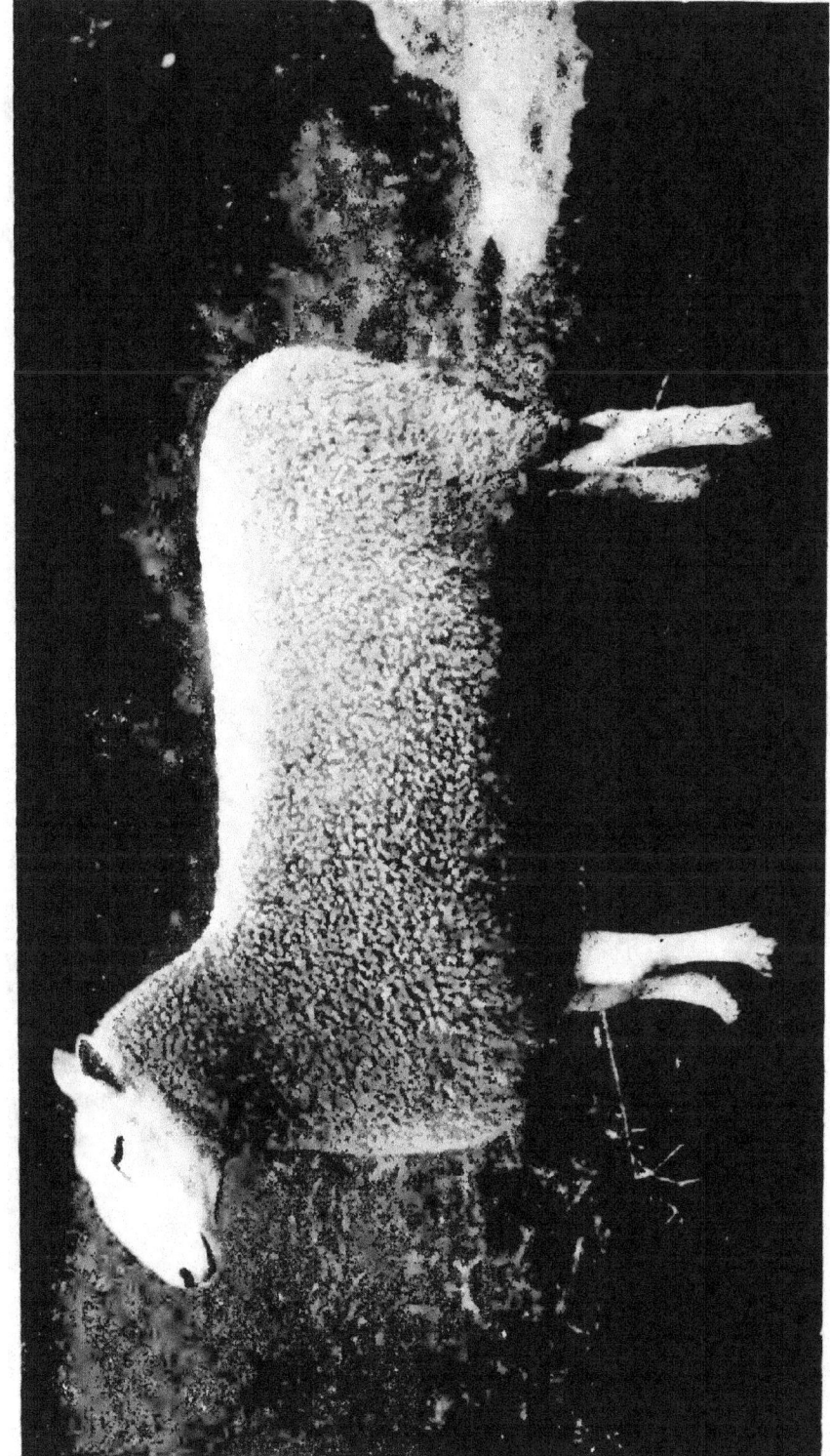

BORDER LEICESTER RAM.

Photo by C. Reid.

LEICESTER SHEARLING RAM.

Photo by Sport and General.

TWO SHEAR HERDWICK RAM.

Photo by *Sport and General.*

DERBYSHIRE GRITSTONE TWO-SHEAR RAM.

LONK SHEARLING RAM.

Photo by *Sport and General*.

SCOTCH HILL, BLACKFACED RAM.

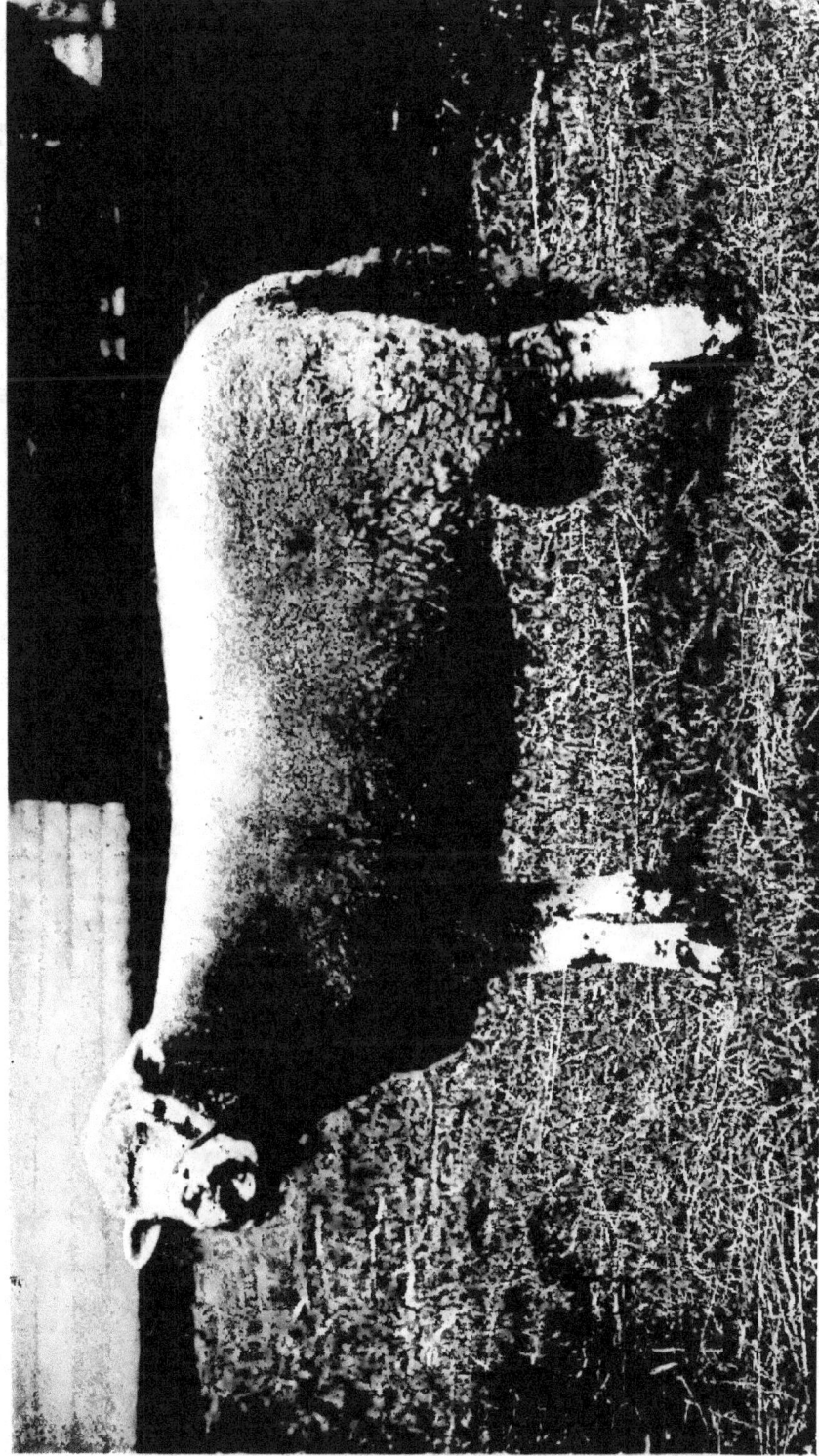

Photo by Sport and General.

[KERRY HILL (WALES) SHEARLING RAM.

Photo by Sport and General

WELSH MOUNTAIN RAM.

DARTMOOR RAM.

Photo by Sport and General.

CHEVIOT RAM.

Photo by C. Reid.

RYELAND SHEARLING RAM.

PEN OF DORSET HORN RAM LAMBS

NORFOLK HORNED RAM

SUFFOLK SHEARLING RAM

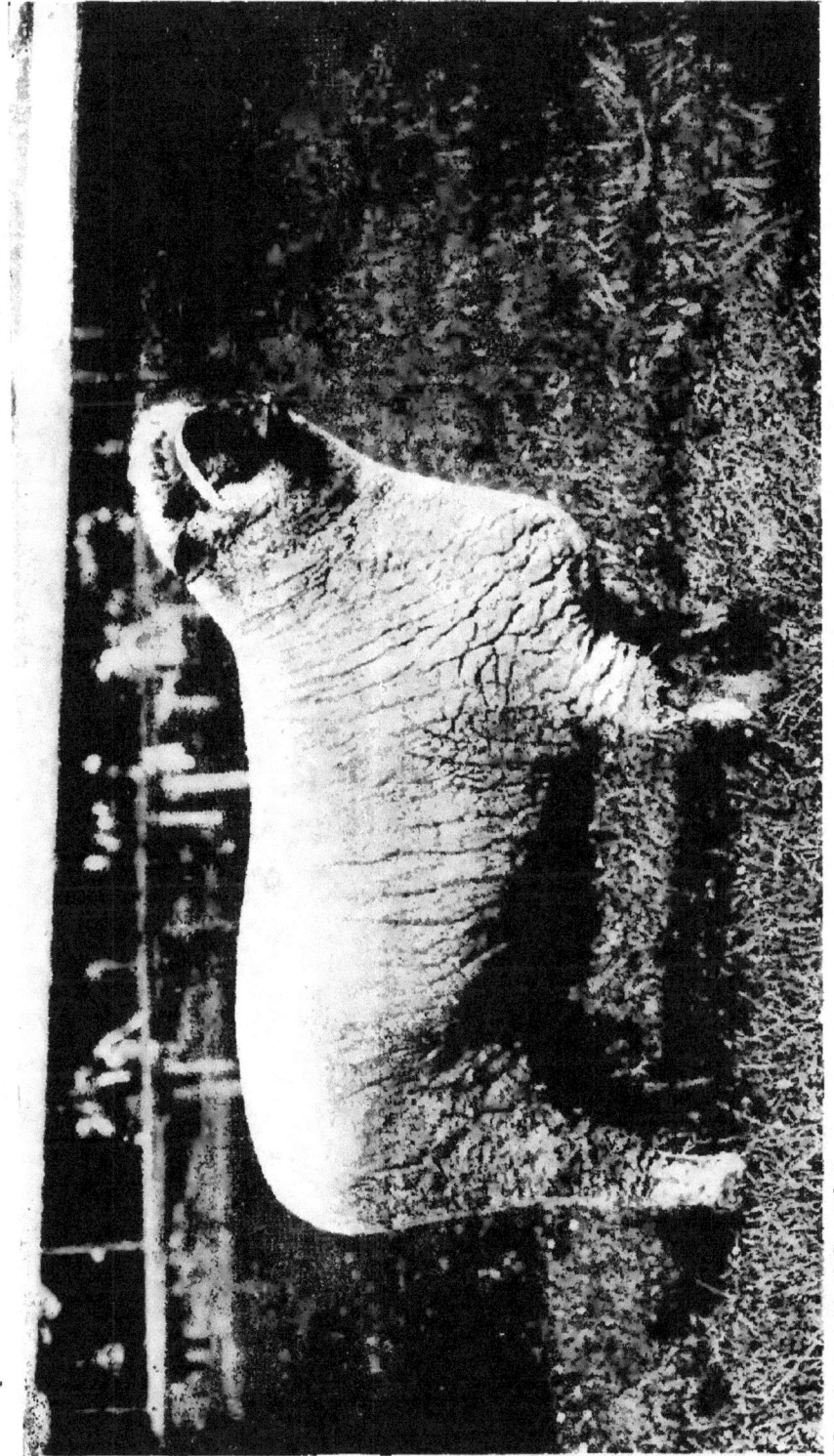

Photo by Sport and General.

OXFORD DOWN SHEARLING RAM.

HAMPSHIRE DOWN RAM LAMB.

Photo by *Sport and General.*

* *

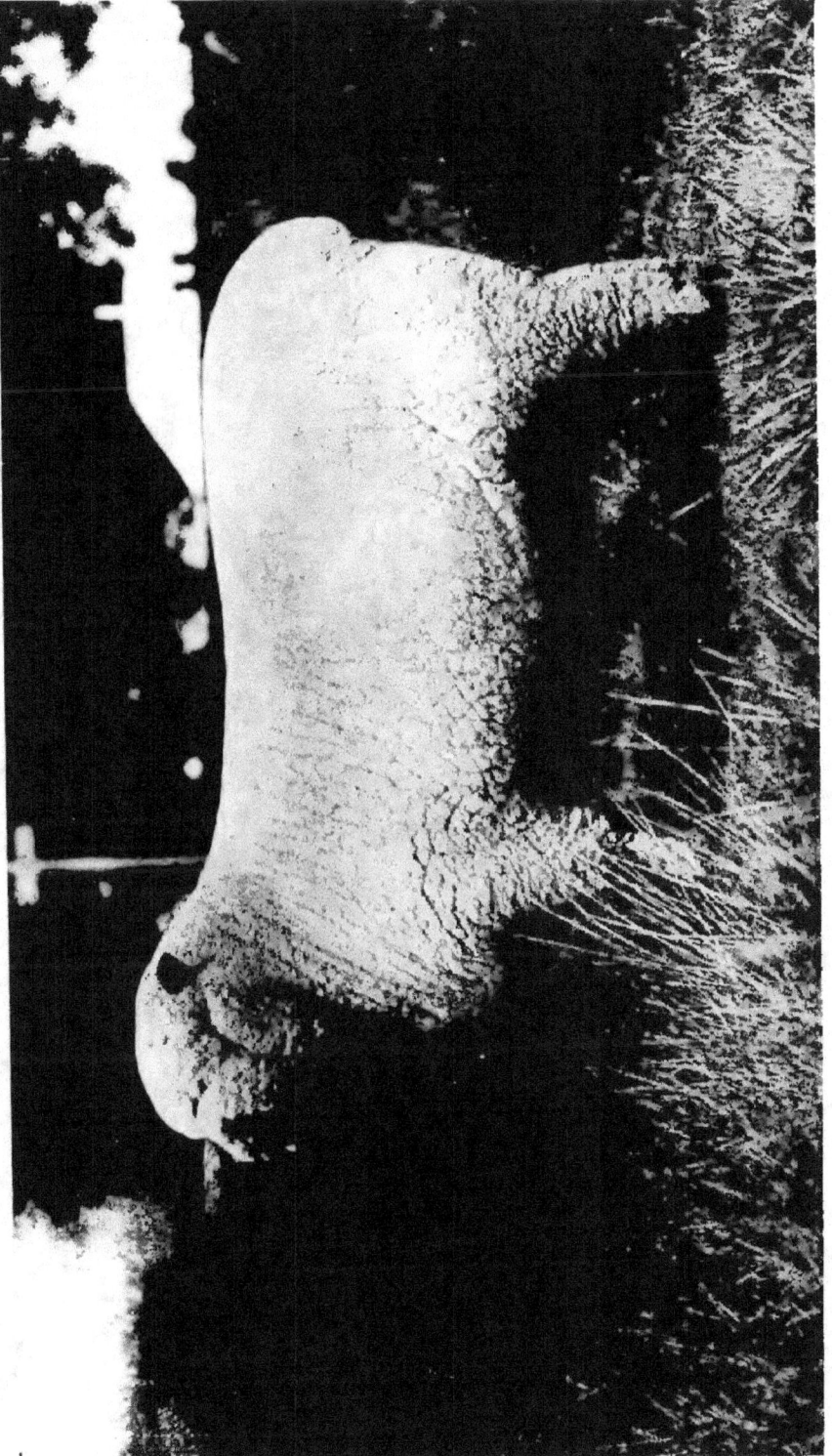

Photo by *Sport and General.*

SHROPSHIRE SHEARLING RAM.

SOUTHDOWN TWO-SHEAR RAM.

Photo by Sport and General.

www.ingramcontent.com/pod-product-compliance
Lightning Source LLC
Chambersburg PA
CBHW081716220526
45468CB00008B/1864